SHARK

Acclaim for *Shark: In Peril in the Sea*

"Owen takes a fascinating look at the biology of sharks, from the smallest (the 19-centimeter-long dwarf lantern shark) to the whale shark, the world's largest fish. He also explores the complex relationship between man and shark. Sure, we eat each other, but the body count is horribly lopsided. While sharks attack only a few dozen people each year, the annual shark catch routinely tips the scale in the hundreds of thousands of tons. *Shark* is a captivating portrait of creatures that have too long been unfairly maligned as malevolent, mindless eating machines."

– *Science News*, (Washington, D.C)

"Overall an excellent account of both the natural history and cultural history of sharks. Having read a few marine natural history books recently I am astounded at how little we really know about what is a fragile environment and its inhabitants. This book will stay on the bookshelf to be used whenever a shark is mentioned to look up the details of that particular shark! Excellent book that dispels the myths of shark phobia, and discusses the IMPORTANCE of Shark's role in the eco system." – *Goodreads.*

"Overfishing and destructive catch methods may have already forced species we didn't even know existed into extinction, and ruined whole ocean habitats that were essential to their survival. We have already proven ourselves capable of ignoring extinctions in our immediate environment. As the oceans are, to a great extent, out of sight, out of mind, it is easy to understand the danger complacency poses for sharks. Owen is correct when he urges a change in attitude. We must protect vulnerable species of shark with the same commitment that we afford whales. Otherwise we may witness another human-induced mass extinction."

– **Kathleen Steele**, *Australian Book Review.*

SHARK

In peril in the sea

DAVID OWEN

Second revised edition

EER

Edward Everett Root, Publishers, Brighton, 2024.

EER
Edward Everett Root, Publishers, Co. Ltd.,
Atlas Chambers, 33 West Street, Brighton, Sussex, BN1 2RE, England.
Full details of our stock-holding overseas agents in America, Australia, China, Europe and Japan, and how to order our books, are given on our website.
www.eerpublishing.com

edwardeverettroot@yahoo.co.uk

We stand with Ukraine!
EER books are **NOT** available for sale in Russia or Belarus.

Shark. In peril in the sea,
first published by Allen & Unwin, Australia, 2009.

This revised edition first published by
Edward Everett Root Publishers Co. Ltd., 2024.

ISBN: 9781915115249 Hardback
ISBN: 9781915115256 Paperback

Internal design by Lisa White. Index by Puddingburn Publishing.
British production by Pageset Ltd., High Wycombe, Buckinghamshire.

For Leisha, Hilton and Larry

CONTENTS

ACKNOWLEDGEMENTS

I would like to convey my thanks to the following who generously provided information and advice for the first edition: John Stevens, CSIRO Marine and Atmospheric Research; Matthew T. McDavitt; Jeffrey Gallant and Chris Harvey-Clark, Greenland Shark and Elasmobranch Education and Research Group. I am also indebted to the work of the late R. Aidan Martin, whose enthusiasm and knowledge was inspirational to me. Needless to say any errors of fact or supposition are mine.

INTRODUCTION

There are many classic shark stories. Here is mine. Early in my research into this book, circa 2005, I keyed into Google the phrase, 'Evolution of the batoids'. Quick as a flash Google shot back, 'Did you mean "evolution of the bastards"?' It became apparent that the language of sharks is not universally known—the skates and rays, close relatives of sharks, are the batoids—and this cyber communication also ironically seemed to confirm our irrational dislike of sharks. In this book's first chapter there is a description of a tragic event, a gruesome death of a surfer, one response to which refers to the sharks involved as 'bastards'. Such reactions are understandable. But they also reaffirm our highly ambivalent attitude towards the natural world. This book is not just about sharks but about our complex association with them over thousands of years, and the grim fact that after well over half a century of intensive marine exploitation and habitat destruction we are rapidly sending many species, both well-known and little-known, towards extinction. To take but one example: in 2020 the International Union for Conservation of Nature (IUCN) listed over 50 per cent of shark and ray species at risk of extinction in the heavily fished Mediterranean Sea.[1]

The unusual terminology of these animals extends to

neoselachian, elasmobranch, chimaera, holocephali, chondrichthyan. Yet these ancient and arcane Greek-root terms are the sum of our modern fears: the fishes that are grouped together because their skeletons are made of cartilage rather than bone. Not only is the formal terminology of the world of sharks confusing to laypersons; even fundamental definitions seem arbitrary. Thus the monumental *The Compact Edition of the Oxford English Dictionary* defines 'selachian' as 'the sharks and their allies', a very broad sweep of the canvas. The chondrichthyans represent no more than five per cent of fish species. Their evolutionary histories are closely intertwined, the skates and rays having evolved as bottom-dwelling 'flat sharks', although some molecular studies suggest that the long-held, science-based evidence for this may be in need of revision. Furthermore, there is much debate and disagreement over the taxonomic structures within the shark, skate and ray orders and families. Again, this is understandable: because cartilage does not fossilise like bone, the meagre shark fossil record is hard to decipher and even with today's sophisticated technologies it is difficult to study the biology and habits of many marine creatures, as their habitat ranges are vast and they live at great depths.

So it is that at both the scientific and popular levels sharks continue to confound our perceptions of them. Proportionately very, very few people study sharks, dive with sharks, or handle sharks in the course of professional or recreational activities—and far fewer still are those injured or killed by sharks. For almost everyone, therefore, shark knowledge derives from aquariums, television documentaries, feature films and YouTube clips, media stories and reportage and of course books, the covers of which all too often feature the hugely gaping mouth of a great white shark—surely the most overused and misleading animal image of all time. But it is that very overuse that is so informative about us. In 2023 the British Library catalogue listed over 2000 books with the word 'shark/s' in their titles. Here is a short selection of these titles; it is a revealing snapshot:

- *Sharks: The silent savages*
- *Sharks: History and biology of the lords of the sea*
- *Sharks and Sardines: Blacks in business in Trinidad and Tobago*
- *The Jaws of Death: Shark as predator, man as prey*
- *Loan Sharks and Their Victims*
- *Sharks and Shipwrecks*
- *The Man Who Rode Sharks*
- *Surprising sharks*
- *Snappy sharks*
- *Sharks in the arts: from feared to revered*
- *Why, why, why are sharks so scary?*
- *Shark drunk: the art of catching a large shark from a tiny rubber dinghy in a big ocean*
- *The shark and the goldfish: positive ways to thrive during waves of change*

The mid-1970s *Jaws* phenomenon, both novel and film, demonised the great white shark (*Carcharodon carcharias*) by cheaply enthralling and horrifying millions of people. It also created such a counter-wave of interest in the elasmobranchs that what is now a major worldwide conservation effort aimed at protecting sharks can in large part be dated back to that negative cultural exploitation of a top-order marine predator. It's most ironic. And fitting, too, that the late Peter Benchley, author of the novel *Jaws*, became a leading figure in the elasmobranch conservation movement. But the complex nature of human–shark interaction is not a modern phenomenon. This book analyses over two thousand years of it. And *Jaws* the story, by tapping so successfully into the fear-driven aspect of that complexity with unexpected results, is an occasionally recurring theme for this book.

Tragically, a far more ancient cultural exploitation of sharks has become one of the greatest threats to the survival of some species in the twenty-first century. During the Chinese Qing Dynasty, from the seventeenth to the early twentieth century, shark fin soup became established as a luxury reserved for emperors. It has ever since retained that high social status, not just because it is protein-rich and has a texture of strand-like silkiness in the mouth, but because it is expensive. The great economic boom in China that began towards the end of the twentieth century put this sought-after delicacy within reach of millions of people, generating unchecked slaughter of sharks at sea, with the living animals, their fins hacked off, often thrown back into the water to sink and die. The organisation Humane Society International estimates that some 72 million sharks are killed annually for shark fins soup, and that a bowl can cost up to $USD100.[2]

Reducing and controlling shark finning has been a major concern of the many environmental entities that have come into being since the 1970s. They are also concerned with an equally wasteful and destructive form of elasmobranch exploitation, one that enters almost every western kitchen and restaurant. Commercial fishing, whether by trawl, net or longline, brings up something in the order of 100 million tonnes of elasmobranchs every year, much of which traditionally was discarded as unwanted bycatch. Waste and cruelty apart, such exploitation practices are unsustainable because of the slow reproductive biology of sharks. Mitigation efforts are making some progress. To take one example: the Marine Stewardship Council, an international non-profit organisation seeking to end overfishing and restore fish stocks, introduced a policy that

> From September 2020, the vessel of any company or fisher convicted of shark finning will not be eligible for MSC certification for at least two years. If evidence of shark finning is detected during an audit or assessment, a fishery will face suspension unless it can show the offending vessel has been expelled.[3]

Ironically, but with its own form of logic, (as explained in a 2015 article) increased finning-related legislation to encourage full utilisation of carcasses

> has seen the market for shark meat expand considerably. In turn, this has led to fishers seeing sharks increasingly as commercial species to be actively targeted, rather than by-catch species landed unintentionally while targeting more valuable species such as tuna or swordfish. The net effect . . . has been to increase fishing pressure on many shark populations, including those whose geographical distance from the end consumer had previously kept them relatively untouched. It has also greatly complicated the task of ensuring that the economic incentives driving the now-global industry do not result in the continued unsustainable utilisation of shark resources.[4]

In 2008 a team of research scientists operating at the fossil-rich Gogo Formation in Western Australia made a unique and instantly famous discovery, a 380-million-year-old female placoderm, an ancient shark fossil, with an embryo attached to it by its umbilical cord. (It was named *Materpiscis attenboroughi* in honour of David Attenborough.) Not only does this confirm that the shark lineage is ancient; it also demonstrates the method of reproduction is very different to that of the bony fishes, the teleosts, which reproduce by the male fertilising eggs ejected by the female into the water. It is not known why elasmobranchs, like mammals, mate and internally fertilise, but one consequence of this method of reproduction is that litters are generally small. Furthermore—again, unlike teleosts—sharks can take many years to become sexually active. Perhaps these life history parameters have always ensured that the elasmobranchs, being the predators and scavengers of the oceans, don't overpopulate. But it makes them particularly threatened by exploitation.

The one thing that everyone knows about sharks is that they have lots of teeth. In some species, there are many hundreds of sharp, multi-shaped teeth in the mouth at any one time. Others have pavement-like crushing dental architecture. As will be described in detail, the sophistication of shark teeth and jaws is astounding. And other shark attributes, such as their numerous sensory systems, and the body streamlining of the fast open-ocean species, have been studied and adapted to various human technological ends. Not so long ago that would have seemed laughable, when sharks were generally ignored or, worse, despised.

Shark intelligence is impossible to understand. Yet the most feared shark seems almost to know us:

> *Carcharodon carcharias* is renowned for his tendency to raise his head above the surface of the sea to observe the

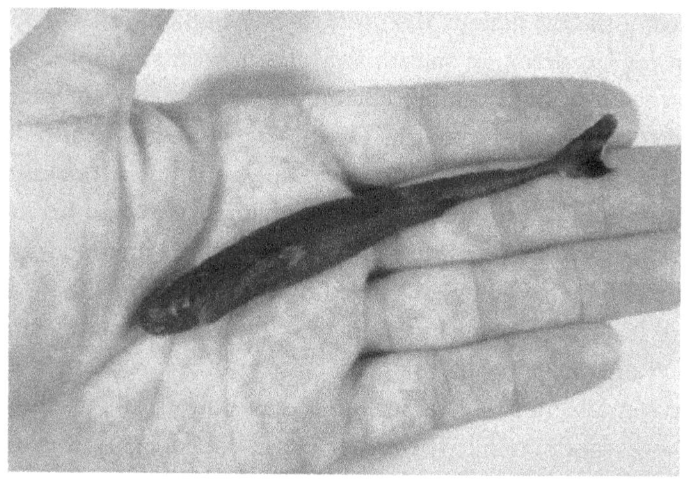

The smallest known elasmobranch is the dwarf lantern shark (Etmopterus perryi), *growing to a maximum of just nineteen centimetres. The largest elasmobranch is the whale shark* (Rhincodon typus), *which may grow to more than fourteen metres in length. This photograph of a stuffed whale shark was taken in Miami, Florida, c. 1912.*[1]

> topside activity of men in boats. This unnerving habit, unique among all shark species, has been well documented by seafarers the world over including those who work on, and cruise, Tasmanian waters. A number of island fishermen have experienced the heart-stopping moment when a white pointer's head has suddenly emerged above the waterline alongside their vessels. All agree that the sharks in question were watching them, even assessing them.[5]

The Chondrichthyes class currently comprises some 1282 species,[6] most of which are small, unobtrusive and harmless. New species are discovered as research into these animals intensifies. The diversity of sharks is bewildering, yet their collective age has meant that each species, no matter how little understood by us, has evolved over millions of years to take up its specific niche in the vastness of the oceans—and, in a few cases, in freshwater systems. The final chapter in this book describes a representative selection of species, from the famous to the innocuous, the common to the rare. Not only is there is no such thing as a boring shark, our own long-term survival, through the health of the seas, relies on them, and the battle to save the planet's sharks has become a deadly race against time.

GLOSSARY

agonistic: competitive aggressive or defensive behaviour, usually within the same species.

ampullae of Lorenzini: electromagnetic sense receptors on the snout.

anterior: the front part.

barbels: tendril-like sense receptors on the head or snout.

basihyal: a structure on the base of the mouth with a suction function.

batoids: skates and rays.

benthic: relating to the bottom of the ocean.

bioluminescence: body light generated by organs.

bivalves: molluscs whose external shell is formed of two halves.

braincase: the single structure enclosing the brain.

cartilage: firm flexible tissue.

caudal fin: tail fin.

caudal peduncle: tail stalk, often with keels.

caudal vertebrae: bony segments of the vertebral column in the tail.

cephalopods: marine molluscs including octopus, squid, nautilus.

cetaceans: aquatic, mostly marine mammals including whales, dolphins, porpoises.

chondrichthyans: fishes with a cartilaginous skeleton.
circumglobal: worldwide within a specific range.
claspers: paired cartilage extensions of the male pelvic fins which transfer sperm to the female cloaca.
cloaca: the opening between the pelvic fins used for reproduction and passing of waste products.
cnidarians: mostly marine animals, including jellyfish, corals, sea anemones.
continental shelf: the coastal seabed to a depth of about 200 metres.
continental slope: the steep sloping seabed beyond the shelf, to about 4000 metres.
copepods: tiny crustaceans making up a large part of the zooplankton.
crocodilians: marine crocodilians are the saltwater crocodiles and caymans.
crustaceans: characterised by jointed limbs, they include crabs, lobsters, shrimp, krill and barnacles.
demersal: living on or near the bed of a deep body of water.
dermal denticles: toothlike placoid scales (dentine enclosing a pulp cavity) embedded in the dermis, the tips of which erupt through the skin.
dermis: connective tissue beneath the epidermis, the outer layer of skin.
diel vertical migration: rising towards the surface to feed at night and returning to depths during the day usually to avoid predators.
echinoderms: exclusively marine animals including starfish, brittle stars, sea cucumbers and sea urchins.
elasmobranchs: sharks, rays and skates.
endemic: found only in a specific area.

endoskeleton: an animal's internal support structure.

endothermic circulation: in some lamnoid sharks, retention of body heat in the blood, elevating the body temperature above that of the water.

epiphysis cerebri: light-detecting gland under the skin on the head, also called the pineal gland.

euphasiids: marine shrimp-like invertebrates, or krill.

fin girdles: the structures that anchor the fins to the body.

fin radials: multiple segmented supporting rods that fan out into the fin base.

fusiform: tapering at each end.

gill arches: structures that support the gill filaments.

gill filaments: tissue structures containing rows of lamellae in which gas exchange takes place.

gill rays: cartilage rods to which the gill arches attach.

heterocercal: unequal in size.

holotype: the single specimen designated as the type for describing a species.

homocercal: similar in size.

ichthyologist: a zoologist who studies fish.

keels: fleshy ridges.

lateral line system: lines of pits, grooves and canals in which are mechanoreceptors sensitive to water movements.

littoral-fringe: occurring near the shore.

monophyletic: an ancestor and its descendants.

morphometrics: the size and shape of an organism.

nares: paired nostril-like holes on the snout and the mouth.

nasoral grooves: furrows that connect the nares to the mouth.

neoselachian: modern (living) species of sharks and rays.

nictitating membrane: a third eyelid which can completely cover the eye.

nudibranchs: soft-bodied marine molluscs without shells.

obligate ram filtering: passive intake of water from which plankton is extracted.

obligate ram ventilation: water enters a swimming shark's mouth and passes over the gills, enabling gaseous exchange.

oophagy: the practice of live young developing within the mother feeding on unfertilised eggs.

oscillatory propulsion: movement through wing-like flapping of the pectoral fins.

otic capsules: paired structures containing the inner ears.

oviparous: laying eggs that develop and hatch independently of the female.

ovoviviparous: producing embryos that develop within eggs in the mother's body, nourished by the yolk.

palaeoichthyology: the study of fish fossils.

pelagic: describing the upper ocean zone.

pharynx: throat region.

philopatry: in migrating animals, the tendency to return to the same area to breed or feed.

phytoplanktons: microscopically small plantlike organisms commonly occurring as bacteria or algae.

pinnipeds: fin-footed semi-aquatic marine mammals, being the seals, sea lions and walruses.

pit organs: sensory organs scattered across the upper body of many elasmobranch species.

polychaetes: segmented, bristled marine worms including lugworms, sandworms and clam worms.

protrusible: capable of being thrust outward/forward.

quadrangular: having four distinct sides.

requiem sharks: also called the whaler sharks, the family Carcharhindae.

respiratory pumping: the ability of many elasmobranch species to pass water over the gills while motionless.

rete mirabile: netlike masses of arteries that capture heat from warmed blood and transfer it to colder blood.
rhomboidal: in the shape of a diamond.
rostrum: a projecting snout.
saddles: coloured blotches on the upper part of the body.
serrated: having sawlike edges.
sirenians: marine mammal herbivores, the manatees and dugongs.
spatial dynamics: the study of geographic movements of an organism.
species: organisms that breed and reproduce with their own kind.
spiracles: remnant gills evolved as paired openings at the front end of the pharynx.
squalene: a liquid organic compound, lighter than water, that is obtained from shark livers and forms the chemical basis of the steroids.
stromatolites: layered structures formed in shallow water by either biological or non-biological processes.
taxonomy: the science and practice of classification, particularly of plants and animals, using specific classification schemes.
terminal: the end part, front or back.
thorax: the region between the head and stomach.
underslung: jaw underneath the head.
ventral: describing the underside.
vertebral centra: strengthened disk-like structures from which the arches of the vertebrae radiate.
viviparous: bearing live young that develop within the mother.
zooplankton: aquatic organisms ranging in size from tiny krill to large jellyfish.

1

SHARK ATTACK

Controversy, Reality, Response

> *The black fin seems graffitied on the [cinema] screen. Like the one drawn on the Amity girl [advertising billboard] it has no depth of field—it looks extraordinary because it's so literal-minded, the opposite of abstract. And we can't help but snigger as this big, black figurative fin sneaks up behind a pretty girl and is noticed by bathers who start yelling and heading for the shore. But first, they stare straight at it. Straight into the camera. For a few seconds, we are the shark, and you know what? These feeble bathers are fair game.*[1]

As this analysis of a scene from the 1975 movie *Jaws* suggests, the relationship between shark and human is a complex one. Make-believe and reality are intertwined as comedy competes with horror, popular entertainment feeds off gruesome tragedy. Like a skull and crossbones, a triangular fin in water is a profoundly evocative symbol. Yet very few of those fins ever 'attack' people.

An incorrect perception of sharks, but one that stubbornly persists, is that they are inherently a danger to human beings. Equally wrong is the belief that large predatory sharks exist in a state of constant hunger and are therefore reflexively conditioned to attack other living things. This is the shark as mindless eating machine, epitomised by an 'infamous' four: the great white shark, the bull shark (*Carcharhinus leucas*) and the tiger shark (*Galeocerdo cuvier*) inshore, and the oceanic whitetip shark (*Carcharhinus longimanus*) the open ocean hunter.

Jacket cover from the 1974 novel Jaws by Peter Benchley. (Used by permission of Doubleday, a division of Random House, Inc.)

The very term 'shark attack' is misleading because it implies that a naturally feeding shark, or a shark displaying instinctive territorial behaviour, is savage and indiscriminate. A child mauled by a family pit bull terrier, however, is by definition attacked. According to wildlife zoologist Dr. David Pemberton:

> There are many reasons to attack. Fighting dogs are trained to fight other dogs and food has nothing to do with it. Animals fight over food, space, dens (which is a form of space) and breeding. Some will fight to defend other animals, one example being a hippo saving an antelope from a crocodile. A leopard seal can kill a person by drowning them then let-

> ting them go; this appears to be space competition. A leopard seal on the other hand can kill or try to kill a person by punching through the ice and grabbing them: this suggests it is after food and sees the human in the light of other prey it takes in this manner, such as emperor penguins. A predator attacks for food and this is clearly proven if the prey is eaten. Sharks killing humans usually eat large amounts of the prey. Pit bull dogs attacking people seldom eat them. The trained killer attack dog loses its urge to keep attacking once the prey is still—hence for many species playing possum is a strategy to stop an attack. The shark does what it is developed to do and that is to survive through eating.[2]

Many shark researchers refer not to attacks but to incidents or interactions, being rare chance encounters with a variety of outcomes, from death to minor lacerations. It is noteworthy that prevailing scientific opinion at the turn of the twentieth century was that sharks were of a cowardly nature and had weak jaws. An eminent American scientist, Dr Frederic Augustus Lucas, whose opinion was sought at the time of a horrifying spate of incidents along the New Jersey coast in 1916, concluded that 'no shark could skin a human leg like a carrot, for the jaws are not powerful enough'[3] —let alone bite through a femur.

A reasonable but unproveable belief is that sharks often mistake their human victims for other forms of prey. While this usefully overturns the notion that sharks target human beings, the reality is that millions of years of evolution have equipped sharks with an array of highly attuned prey-identifying sensory capabilities, including electrical detectors, acute senses of smell and hearing and exceptional eyesight in some species. When a shark is uncertain about a potential food source, it is likely to purposefully approach and investigate, nudging or biting. Furthermore, the feeding regimes of the named four species are not exclusively predatory, because they all scavenge. Thus an object bobbing on the surface, be it a surfer, a seabird, a chunk of whale blubber, a person or dog swimming, or a corpse in a river or far out sea, is indeed fair game.

Other explanations for why sharks mistake human beings for animal marine prey that is part of their normal diet are that the flash of jewellery or other bright objects, and the pale soles of feet, resemble shoaling fish. Likewise, the erratic splashing of a bather or swimming style of a dog might effectively mimic marine creatures in distress, or resemble some other underwater prey sound, although it is just as likely that surface turbulence attracts sharks regardless of how it is made (sharks are known to bite boat propellors).

The most compelling case of mistaken identity, widely believed to be true, relates to great white sharks and their primary prey, pinnipeds (walruses and seals). The belief is that a person in a black wetsuit on a board, when seen from under water backlit against sky, sufficiently resembles a pinniped to become a target. Respected ichthyologist R. Aidan Martin rejected the notion of mistaken identity, believing it to be 'completely false', based on his observations of great white shark hunting behaviour at South Africa's Seal Island. The great white is an ambush predator which typically stalks from below then launches itself up at its chosen target. Martin observed many attacks on seals, in which 'the sharks would rocket to the surface and pulverize their prey with incredible force'. By contrast, humans are frequently investigated with 'leisurely or undramatic behavior' in order to determine their edibility, and this may include an exploratory bite.[4] The fact that a great white shark has excellent eyesight should also be considered.

The case against mistaken identity is strengthened when considering that a shark, or any predator for that matter, will not expend unnecessary energy against a motionless or slow-moving source of potential prey. In 1987 Craig Rogers, a 40-year-old landscape contractor, was surfing with a friend off Santa Cruz, California. As he sat on his board, his legs dangling, a movement made him look down and he saw a huge shark biting his board, next to his hand. It had come up so stealthily that Rogers hadn't heard it, and then, 'he watched the shark gently release his board and sink like a submarine.' Furthermore,

> During its gentle contact with Rogers' surfboard, the shark left its 'calling card' in the form of two whole teeth. In addition, the impressions left by the shark's teeth were easily identifiable by their spacing. This permitted a reliable size estimate for the attacking White Shark at 5.7 metres in length. The surfboard, with accompanying White Shark teeth, is currently in a surfing museum in Santa Cruz.[5]

Set against this are eyewitness accounts describing the ferocity of attacks on people by sharks. In April 2008 Dave Martin, a member of a triathlon training group, was killed by a great white while he was swimming off southern California's Tide Beach. The shark 'struck around 7 am, charging at Martin from below and lifting him vertically out of the water, both legs in its jaws, its serrated teeth slicing deep, fatal gashes. "They saw him come up out of the water, scream 'shark', flail his arms and go back under", said Rob Hill, a member of the Triathlon Club of San Diego, who was running along the beach when the attack happened.'[6]

Debatable, if not strictly for movie fans, is the concept of the rogue shark, the individual that develops a taste for human flesh. Before the introduction of shark netting, the coastlines of Australia, the United States and South Africa all occasionally experienced so-called cluster attacks—repeated attacks in a specific geographical location in relatively concentrated periods of time. They were rare and all but ceased after protection measures were introduced at swimming beaches before the mid-twentieth century. What might explain cluster attacks? It could be that an ageing, sick, wounded or mentally disoriented shark is reduced to hunting and scavenging in a localised area. This is not *Jaws*-style rogue behaviour. Similarly, where there is no anti-shark protection and sharks feed on human beings, for example high seas shipwreck victims, or bathers or corpses in rivers or estuaries, this is not rogue behaviour but a combination of predatory and scavenging behaviour. Bull sharks, also known as river whalers (Zambezi sharks in Africa), were once known to congregate some 90 kilometres inland at the site of the Ramornie Meat Works on the bank of the Orara River in northern New South Wales, in order to scavenge off disposed offal. And, like crocodiles, sharks feed off corpses ritually consigned to the holy waters of India's rivers and Lake Nicaragua. And, like seagulls, they follow ships for their refuse tossed overboard.

Shark attacks are statistically very rare and usually not fatal. Since 1958, the American Elasmobranch Society and the Florida Museum of Natural History have been collating the International Shark Attack File (ISAF). The ISAF has records of every known shark attack dating back some 500 years, which in 2023 totals more than 6800 individual investigations. Since the beginning of the twenty-first century, the ISAF has recorded a mean average of five known unprovoked fatalities per year.[7]

Unprovoked bites are defined by the ISAF as *incidents in which 'a bite on a live human occurs in the shark's natural habitat with no human provocation of the shark.' Provoked bites occur when 'a human initiates interaction with a shark in some way,' examples of which are people being bitten attempting to feed or touch or unhook a shark.*[8]

What can't be known is the number of unreported attacks, particularly in developing nations and remote regions of the world. Where there are such studies, they are informative. To take one example: a study of attacks in the rivers of southern Iran documented 34 cases over nearly half a century (1941–1985), half of which were fatal. The number of attacks decreased as the introduction of piped water meant that fewer people used the rivers for water supplies.[9]

ISAF statistics in countries where human activity in shark waters is regular confirm the low rate of shark–human interactions. In 2022 the ISAF investigated 108 'alleged incidents of shark–human interaction worldwide'. Fifty-seven were classified as unprovoked. Five of the unprovoked attacks were fatal (of a total of nine shark-related fatalities). As a viable comparison, between 2004 and 2007 in the US there were 15 shark attack fatalities and 127 dog attack fatalities.[10]

The ISAF also compiles statistics on shark species involved in attacks—with, however, the proviso that such data should be used with caution, because species identification by victims and witnesses is often wrong. Total recorded species attacks between 1580 and 2008, including fatalities and non-fatalities, were: great white shark, 451; tiger shark, 158; bull shark, 120; sand tiger shark/grey nurse shark (*Carcharias taurus*), 75; blacktip shark (*Carcharhinus limbatus*), 41; hammerhead shark (*Sphyrna spp.*), 41; blue shark (*Prionace glauca*), 41. About 35 other species were also cited, with such minor infractions that the records become meaningless, even comic. As an example the basking shark (*Cetorhinus maximus*) has been implicated in two 'boat attacks' and wobbegongs (*Orectolobus spp.*) implicated in two non-fatal unprovoked attacks. (Interestingly, though, there are 21 records of wobbegongs engaging in provoked attacks, suggesting a lack of caution or respect on the part of the attacked.) Needless to say, the number of humans consumed by sharks increases greatly when factoring in victims of air and sea disasters.

ISAF data also reveal distinct patterns to shark attacks, the most common being a 'hit and run', where a shark bites once and does not return to the victim. This behaviour could result from any number of factors. 'Bump and bite' attacks are often preceded by the shark circling its victim, which is a clear demonstration of intent on the part of the animal. The shark is probably assessing the potential resilience and nutritional appeal of the target. A 'sneak' attack occurs without warning and seems to be a premeditated ambush. In the latter two cases the shark will often repeatedly attack, resulting in severe injury or death to the victim.

The ISAF asks victims and/or witnesses to complete its Shark Attack Questionnaire, which is designed to help identify the likely species involved. Respondents are asked to choose one action out of each of the following lists:

Shark behavior prior to initial strike
Circling victim
Following victim closely
Shark in position between victim and barrier or obstacle/beach/reef/boat, etc.
Shark not seen at all prior to contact with victim
Straight and direct approach to victim

Straight and direct approach to victim, passed close by other(s) in water
Swimming erratically
Swimming normally
Behavior unknown
Other (please describe)

Shark behavior at time of initial strike
Attack did not occur in water
Shark did not contact victim
Minimum of turmoil, victim initially unaware of situation
Sudden violent interaction between shark and victim
Behavior unknown
Other (please describe)

Shark behavior during subsequent strikes
Attack did not occur in water
Shark made only one strike
Shark made multiple/repeated deliberate strikes
Frenzied behavior
Released initial hold, quickly bit victim again
Behavior unknown
Other (please describe)

Shark behavior after final strike
Attack did not occur in water
Shark remained attached to victim and had to be forcibly removed
Shark remained in immediate area of attack
Shark followed victim/rescuers towards shore
Shark seen to leave area of attack
Shark not seen after final strike

Shark remained attached to victim after final strike, released hold without use of force by victim/rescuer
Behavior unknown.
Other (please describe)[11]

A description of the feeding method of predatory sharks illustrates just why attacks can be so destructive:

> Their initial attack on a victim too large to be swallowed whole is usually aimed simply at taking out a mouthful of flesh . . . In the case of an attack on a [human], this often results in a leg being take clean off, especially if the bite goes through a knee joint, or the removal of a huge quantity of muscle from calf, thigh or buttock, leading to fatal haemorrhaging from ruptured arteries. In many cases, the flesh is effectively stripped from the tibia, fibula and sometimes the femur as well when the victim struggles to get free and the shark fights to make off with its pound of flesh.[12]

The United States regularly records most unprovoked shark bites. Australian fatalities have averaged one per year over the last sixty years,[13] although in the first seven years of the twenty-first century there were eleven shark attack fatalities in Australian waters, including two river and canal bull shark fatalities and one near Neptune Island, 70 kilometres off the coast of South Australia (a great white shark congregation zone because of the presence of breeding fur seals, sea lions and seabirds).

The temperate, prey-rich waters of South and Western Australia are particularly favoured by great whites. Australian shark trackers are aware that great whites travel along these coastlines to pupping grounds in South Australia's Spencer Gulf to give birth. They also travel west along the Western Australian coastline into the Indian Ocean. It's not surprising therefore that many surfers in these regions consider sharks to be their foremost occupational hazard.

In 2004 a well-known surfer was killed by a pair of great white sharks off a popular Western Australian beach: the locality and nature of the attack, and the victim's high profile as a well-known surfer combined to generate horror headlines across the country. A written tribute in a surfing magazine evokes something of the psychological complexities associated with such grim fatalities:

> While surfing Noisys near lefthanders in Gracetown on Saturday 31-year-old Brad Smith from Rockingham lost his life doing what he loved best, surfing. He passed away before reaching the shore. Police have closed beaches in the region . . . His friends assembled in the surf beach car park at Secret Harbour last night, the stories and facts came out about a man who loved, life, surfing and his mates . . . A man that would stick up for his mates and was as Australian as Ayers Rock. A hard tough man that called a spade a spade and was scared of nothing . . . How tough was this legend? Friends were describing how he fought the first shark and refused to give in, punching it in the head continuously and giving it the fight of its life which unfortunately cost him his. But remember this, it took two of the Bastards to take him down and he went down with a fight. The second shark leaped out of the water in flight and was also believed to be a White Pointer, one shark was believed to be 3 metres in length while the other was around 5 metres . . . Rest in peace and hope you catch up with Simmo and the rest of the crew and have a great time wherever you are. From all the lads in Rocko and those who have had the pleasure to have met you and surf with you [signed] Mario 'Marz' Vassallo.[14]

Eighteen months later, on Australia's east coast, another fatality created a media sensation through its own tragic circumstances. Twenty-one-year-old Sarah Whiley was holidaying with a church group at North Stradbroke Island, Queensland, when she was attacked at Amity Point Beach by up to three bull sharks. Both of her arms were bitten off, her legs and torso severely mauled and she bled to death while being airlifted to a Brisbane hospital. According to a Queensland police officer, 'She was swimming with three other friends. She went down under the water and screamed "Shark!" And, of course, people at the time thought she was only joking until they saw the blood.'[15] One eyewitness said that she had been swimming about 50 metres away from other people. Another local said that she had been swimming with her dog, which left the water and ran to its home in a 'frenzied' state and had to be locked up.

Amity Point Beach has shark protection drumlines set with baited hooks beyond the beach breakers. However, Sarah Whiley had been swimming in an adjacent channel 'that locals avoided because they feared shark attacks'.[16] This is a purpose-built boat channel, its water on that day described as 'very murky, dirty in fact'.[17] A local fisherman said, 'We've been waiting for this for a long time. We've always thought someone was going to be taken here. I'm a crabber and at this time of the year massive bull sharks come over the [sand] bar. It's nothing to see 10 or a dozen bull sharks under our boat when we are crabbing and they are really aggressive—they are not like normal sharks.'[18]

Valerie Taylor began working with sharks in the 1960s and is famed for her underwater photography and her association with classic shark films including *Blue Water, White* Death and *Jaws*. She commented on the Amity Point attack:

'The sharks would come up that shipping channel, boating channel, to the beach on a rising tide, late afternoon, looking for any food that might wash off the beach and nature intends them to do this.'[19] The extent of the victim's wounds would suggest that an initial 'bump and bite' attack had led to what Taylor called a 'feeding pattern'.

The locations of these two examples—a popular surfing break, a protected beach frequented by tourists—invite the question of why they, and others like them, haven't experienced more such attacks. Furthermore, surfers, kayakers and windsurfers operate well beyond protected beaches, and countless numbers of people swim in unprotected waters that might be deemed risky. Shark experts are not joking when they say that humans are too bony and fat-free for the liking of sharks.

'Big bite: Snr-Sgt Jason Elmer with Hannah's surfboard.' This caption accompanied the image of teenage Hannah Mighall's surfboard. She was repeatedly attacked by a 5-metre great white shark on the east coast of Tasmania, in clear shallow water. Her cousin Syb Mundy wrestled and punched the shark away. Hannah made a full recovery. (Courtesy of *The Mercury*, 13 January 2009)

A horrifying but not fatal attack near Eden south of Sydney in January 2007 resulted in unlikely comedic headlines: 'Shark survivor in a media feeding frenzy'; 'Fish-food fears fuel dad's fight for life; 'From the jaws of a shark comes a ripping yarn'; and, neatly summing up the public's relentless appetite for fear with a happy ending: 'When Eric Nerhus escaped from the jaws of a great white shark, he joined an exclusive club: the people who have tangled with the planet's most dangerous creatures—and yet lived to tell the tale'.[20]

Nerhus, an abalone diver, was operating in murky water off Cape Howe near a seal colony when the shark, about three metres in length, attacked without warning. Nerhus later described the incident: 'I went straight into its mouth. My shoulders, my head and one arm went straight down into its throat. I could feel its teeth crunching down on my weight vest . . . I've never felt fear like it until I was inside those jaws.'[21] The shark then began to shake him from side to side, in the typical manner designed to shear and tear flesh. With his free hand Nerhus used his abalone knife to feel for an eye socket, which he jabbed repeatedly. The shark let him go and he swam to the surface, where his son Mark was waiting in their boat. Other abalone divers in the area rushed to help. After an hour-long return trip to shore he was flown to Wollongong Hospital and treated for bites that required 75 stitches. A fellow abalone diver told a reporter, 'I have had close encounters several times. I haven't seen a shark in that particular area, but that's where they are, because two oceans meet, the Tasman and the Pacific, and there's an upswell. This is black water.'[22]

A similar act of desperate self-defence played out in clear shallow water off Middleton Beach in the south of Western Australia in May 2008. Jason Cull, 37, was swimming about 80 metres from shore when a dark shape that he thought was a dolphin approached him, but,

> It was much bigger than a dolphin when it came up. It banged straight into me. I realised what it was . . . I sort of punched it, and it grabbed me by the leg and dragged me under the water. I just remember being dragged backwards underwater. I felt along it. I found its eye and I poked it in the eye, and that's when it let go.[23]

Two features of this attack are noteworthy. The first is that the shark was a great white of about four or five metres in length and it was accompanied by two others, yet despite the victim's blood loss he was not then subject to a group attack. Instead the sharks were seen to head towards other swimmers. The second is the heroics of Joanne Lucas, a local surfer:

> This is a woman who is 50, and about four foot eight, five foot, not much of her, strong, very strong, a good athlete for the [surf] club. Mother of three, and without any hesitation, knowing there were sharks in the vicinity, swam right up . . . She got hold of the injured swimmer and brought him back to shore. There was a fairly comprehensive mauling of his left leg, lost a lot of his calf, severe lacerations.[24]

In 2009 Australia experienced its own 'Summer of the Shark', a sharp increase in attacks. The phrase had come into being in the United States in the summer of 2001, as a cover story of *Time* magazine capitalising on a frenzy:

> The media coverage was prompted by a bull shark biting off the arm of an 8-year-old boy on a Florida beach July 6, 2001. Overnight, shark bites and sightings became major international news, triggering countless TV news reports and front-page stories and culminating in the Weekly World News tabloid declaring: 'Castro trained killer sharks to attack U.S.'[25]

In fact there was no increase in shark attacks in the United States that year (and fewer shark attacks worldwide), but significant perception negativity was being done to sharks, until the media value of the story came to an abrupt end on 11 September. Australia's 2009 Summer of the Shark, while having more of a basis in reality, with seven non-fatal attacks along the east coast over a seven-week period, nonetheless generated massive media coverage of the kind usually associated with national tragedies.

The irresistible attraction to the media included this headline from the UK *Guardian*: 'Monster munch: three shark attacks in 24 hours throw Australia into Jaws panic'.[26] Much was made of the locations of two of the attacks—at iconic Bondi Beach, widely reported as the first attack there in 80 years, and in Sydney Harbour, the attack there described as within sight of the Sydney Opera House. More meaningful was the debate that ensued. Shark researchers believed that increased numbers of school fish such as salmon and kingfish—their numbers protected by controls on their commercial harvesting—had attracted sharks which might not otherwise be so plentiful in those areas. Other interested parties, including commercial shark fishermen, cited the attacks as proof of a rise in shark numbers to plague proportions, blaming state government protection of shark species considered threatened, especially *Carcharodon carcharias*, the great white shark.

Every shark attack rapidly clears the water—that is understandable. But when it just as rapidly clears the front page, tragedy becomes morbid fascination. As famed Australian shark attack researcher Victor Coppleson noted, shark attacks are 'grotesquely spectacular', and he recognised the impact of grimly descriptive reportage, which he deployed in his account of a 1947 attack at Port Macquarie on the New South Wales coast:

This 1950s photograph was taken off North Bondi Beach, Australia. In 2009 the first attack at Bondi for 80 years became part of an Australian 'summer of the shark' media phenomenon. (From *Shark Attack* by V.M. Coppleson, Angus & Robertson, 1958)

> The boys were swimming in front of their home at The Hatch, 12 miles from the river's mouth. Suddenly, Rupert, 13 years, screamed. There was a swirl of water, and he staggered bleeding towards the shore. Almost immediately there was another high-pitched scream from 12-year-old Edwin, who disappeared below the surface. As soon as he reappeared his elder brother Stanley grabbed him and tried to pull him from the grip of a shark. Suddenly he came free. His leg had been taken off at the knee. On the beach young Edwin died in his brother's arms. In the meantime Rupert had left the water with blood pouring from a gaping wound with the flesh almost completely stripped from above the knee to six inches below the kneecap. Rupert recovered from his severe injuries . . .[27]

Individual and community responses to shark attacks take many forms. During the New Jersey panic of 1916, when four people were killed and a fifth badly injured over a ten-day period by what was thought to be a lone juvenile great white shark, men in boats dynamited the waters. In contrast, after Brad Smith's death in 2004, Western Australian fisheries authorities used a boat and helicopter to try and locate the sharks in order to drive them into deep waters. This was because great white sharks are a protected species—although in this instance the officers were granted authority to kill them—and also because Smith's brother Stephen reportedly requested that the sharks not be killed in 'senseless revenge'.[28] (The sharks were not seen again.) A witness to the attack at Amity Point 'ran up and down the beach screaming at people to get out of the water, and was on the mobile to Triple-0 [emergency number] . . . She was pretty hysterical'.[29] Eric Nerhus, his goggles crushed and his oxygen regulator knocked out of his mouth, estimated that he spent two minutes in his shark's mouth, desperately stabbing at it with his abalone knife.

Not every shark encounter meets with sympathy. In January 2008 *The Australian* was one of many newspapers to dramatise an incident off the east coast. Its headline, 'Shark attacks fisherman off Gold Coast', introduced a vivid lead paragraph in bold print: 'A shark's jaws had latched so tightly onto a man's leg aboard a fishing boat today that its head had to be cut off to free him'. The three-metre shark had bitten the tuna boat fisherman when he accidentally stood on its tail, and he had to be winched into a rescue helicopter in 'a tricky operation, carried out in three- to five-metre swells . . . the victim was lucky to be alive because the bite narrowly missed major blood vessels and arteries at the back of the knee'.[30] Two days later this letter appeared in Melbourne's

The Age newspaper: 'A man puts a bloody great hook into the mouth of a shark, drags it through the water with every intent of killing it, and hauls it out of the water, probably with an even bigger gaff hook. When the shark has the impertinence to fight back, we have a million news stories about a "shark attack". I'd call it just deserts.'[31]

Practical responses to shark attacks are obvious enough. After the 'Black December' of 1957 on South Africa's Natal coast, when seven Zambezi (bull) shark attacks included five fatalities, and devastated that province's tourist economy, its authorities formed the Natal Sharks Board and laid over 40 kilometres of nets at 45 beaches.

The 1945 sinking of the warship USS *Indianapolis* led to the US Navy embarking on a shark-repellent research program (which ultimately resulted in the establishment of the International Shark Attack File). The cruiser had delivered the Hiroshima atomic bomb to the Tinian Island airfield in the Philippine Sea, and was subsequently torpedoed by a Japanese submarine, leaving 900 survivors helpless in the water and undiscovered for four days. Just 316 were rescued. Many had been killed by sharks. According to one survivor: 'The day wore on and the sharks were around, hundreds of them. You'd hear guys scream, especially late in the afternoon. Seemed like the sharks were the worst late in the afternoon than they were during the day. Then they fed at night too. Everything would be quiet and then you'd hear somebody scream and you knew a shark had got him.' [32]As will be described later in this book, the fate of the USS *Indianapolis* plays a significant role in the movie *Jaws*.

A later chapter discusses sharks and conservation, two aspects of which are usefully foreshadowed here. The first relates to oceanic water health. In August 2023 a UK newspaper headline ominously, funnily, or stupidly declared: 'Just when New Yorkers thought it was safe to go back into the water . . .'[33] The article reported on a spate of bathers being bitten by sharks at popular Rockaway Beach. 'Police attempted to reassure locals and holidaymakers that attacks from New York's resident tiger sharks are "extremely rare" but their frequency has increased in recent years".'[34] Helicopters searched for the sharks. The apparent increase in these shark numbers probably results from 'ongoing work to clean the waters around Long Island . . . excellent resource management strategies that have increased not only the shark populations somewhat modestly but also their [bunker fish] prey.'[35] It was also described as a' success story for conservation'. Ascribing increased shark 'attacks' on humans to conservation efforts is truly ironic.

The second aspect is about sharks and climate change. An alarming August 2023 headline, also in the UK *Daily Telegraph* – 'Sharks "angrier than ever" as oceans warm' – quoted numerous oceanographic and climate experts warning of 'dire consequences for marine life' as ocean temperatures reach the highest level on record. Marine heatwaves doubled in frequency and duration between 1982 and 2016. The article stated that 'predatory animals such as sharks can become aggressive as they get confused in hotter conditions', having 'a very narrow tolerance for temperature changes.' These rising temperatures are also predicted to increase the numbers of sharks in waters around Britain, as they move away from southern European waters. 'Angry' sharks and more sharks suggest increased interaction with people in those waters.[36]

Shark repellents are popular. A 2004 patent application lodged with the World Intellectual Property Organization described an 'electric field shark repellent wetsuit' with 'an electroactive material integrated into the clothing, at least one electrode . . . the electroactive material is adapted to release electrical impulses into the water . . . to generate an electrical field around the clothing'.[37]

A company manufacturing various marine products, carrying the logo 'Defence Recognised Supplier—Australian Defence Industry' with a 'Nato stock number' trademarked a 'Shark Shield':

> The Shark Shield . . . generates an electrical field . . . The field is projected from electrodes in the tail of the unit that trail behind the diver. This creates an elliptical field that surrounds the user with a shark safe zone that is up to eight metres in diameter . . . [the field] is detected by the shark through its sensory receptors, known as the Ampullae of Lorenzini, found on the snouts of sharks. Once detected by the shark's sensors, the field causes muscular spasms that result in the shark being deterred from the area . . . there is no lasting detrimental effect to the shark. The unique and unfamiliar pulsing sensation emitted by the Shark Shield does not replicate that given off by a fish or seal and hence it is not recognized by a shark, as a desirable or attractive stimulus. The initial discomfort increases if the sharks approach the transmitter until it becomes intolerable. The sharks then veer away and usually leave the immediate area.[38]

Another company copyrighted and patented a graphic called 'Shark Camo', claiming it to be used by thousands of surfers worldwide. It was tested at numerous locations including in the 'ring of death' waters of South Africa's Seal Island, the protective described as

> . . . a specially designed sticker that is placed on the underside of surfboards (and other water craft) to repel sharks and

> help prevent mistaken identity shark attacks. Gives peace of mind while you're in the water for under 50 bucks. The patented, striped 'fingerprint' pattern mimics the colouration of a number of fish which are not prey for sharks. These fish include the remora, pilot fish, sea snake and lion fish. Scientists agree that sharks use vision as their primary sense when within striking distance of potential prey and so this sticker, placed on the underneath of surfboards, helps repel them from executing a mistaken identity or 'bump and bite' curiosity attack.[39]

For those unconvinced of the efficacy of electrically charged wetsuits or stripy decals, there is always the 'common sense' approach. Numerous shark do's and don'ts can be inferred from the circumstances of the attacks described above and the testimony of eyewitnesses: don't swim in murky water or river mouths; don't swim in certain areas late in the afternoon; don't swim while bleeding; don't swim near fishermen or feeding seabirds, or near steep drop-offs where shoal fish feed and in turn are fed upon by sharks; be wary of diving near seal colonies. Local knowledge and experience are especially important. To take but one example, the Iranian rivers researchers deduced that sharks became more prevalent 'from July to September when freshwater flow was at a minimum and tidal penetration of salt water at its highest'.[40] Some activities are reckless. There was a time when Durban surf fishermen paid surfers to hitch their baits to their bathers, paddle out beyond the breakers and drop the baited hook into what was, literally, shark-infested waters, due to the presence of a whaling station. Invariably, the fisher would have hooked a decent-sized shark before the surfer had caught a wave back to shore.

Ocean Guardian[41] describes itself as 'the world's leading

shark deterrent technology company. Since 2001 the company's Shark Shield Technology has protected tens of thousands of ocean users and prevented the unnecessary killing of marine life by environmentally destructive shark nets, drumlines and culls. Independently proven and tested, Ocean Guardian's Shark Shield Technology is backed by multiple peer-reviewed published research papers . . . Sharks have short-range electrical receptors in their snouts used for finding food. Shark Shield Technology is used to create a powerful three-dimensional electrical field which causes spasms in these sensitive receptors turning sharks away. There are no known harmful effects on sharks or humans.

Surfers, sea-kayakers and divers generally have respect for the unpredictable element they love and its inhabitants. Their advice might be punchy, but it's practical, as this advice for kayakers demonstrates:

> Bull Sharks are just another reason to: learn to roll and not wet exit in the surf zone; paddle offshore instead of harbours, rivers and estuaries. Remember, no one dies from the loss of an arm or leg. They die from blood loss, exposure and drowning. If you get munched, STOP the bleeding as a priority. Direct pressure, compression bandage or tourniquet. Whatever suits you and your training. Me, I'm using the paddle leash. Tourniquet pressure necrosis is not an issue to me if the limb is missing and I am likely to bleed to death at sea.[42]

Just as it would be risky to have a picnic in an enclosed field occupied by a bull, care should be taken in choosing where and when to enter the water. The power and unpredictability of sharks is undeniable—and, as a later chapter shows, those attributes have in the past also engendered great respect.

2
FATHOMING THE SHARK

Evolution, Classification

> *It shows that Steve came over the top of the ray and the tail came up, and spiked him here [in the chest], and he pulled it out and the next minute he's gone. That was it. The cameraman had to shut down.*[1]

This was how a close friend of naturalist and adventurer Steve Irwin, known to the world as Australia's 'Crocodile Hunter', described the camera footage of Irwin's death on 4 September 2006. Irwin had been making a television documentary about dangerous sea creatures and was snorkelling off the far north Queensland coast when he was killed by a large stingray. His international profile and dramatic death ensured massive media coverage, garnering spontaneous tributes from popstars, politicians and members of the public alike. Irwin's death is the sort likely to pass into folklore, or myth, like that of Odysseus, 'who had passed through countless woes of the sea in his laborious adventures, the grievous Sting-ray slew with one blow'.[2]

Had Irwin died in a car crash, or of a more conventional heart attack, public reaction would surely have been less frenzied. The dramatic and unusual nature of the death made it a huge media story. It certainly was unusual but it was not unnatural. In clear shallow water, Irwin and his cameraman had first followed and then swum alongside a large ray, weighing as much as 100 kilograms. It was probably a cowtail ray (*Pastinachus atrus*). Irwin swam close enough for the animal to feel threatened and it defended itself instinctively, whipping around to confront the large intruder and, like a scorpion, arcing up its long tail. It then rammed the tail's poison-coated, 20-centimetre-long barbed spine into his heart. An electrosensitive defence mechanism perfected millions of years ago had locked onto the strong electromagnetic pulses generated by Irwin's beating heart.

The ray no doubt then swam on to rejoin its companions and is probably still alive and well. At such a size it has few predators, and rays have been known to live for up to 50 years. It would also soon have grown a new tail spine. Little known is that a stingray spine is a highly modified tooth. Equally little known is that technically Steve Irwin was attacked and killed by a shark, because rays are described as 'flattened shark derivatives'[3] or, more colloquially, as sharks' 'pancake cousins'.[4] Both descriptions reflect the rays' evolutionary divergence away from the classic torpedo (fusiform) shape as they and the skates adapted to life as bottom dwellers, or 'benthic specialists'.[5] Although recent molecular studies have cast doubt on the closeness of the evolutionary relationship, indicating that the batoids may be a separate monophyletic elasmobranch group, they share many characteristics differentiating them from the teleosts.

Strange as it may seem, shark-shaped sharks are in fact a minority of their biological class, the cartilaginous fishes, being outnumbered in species count by the batoids—the rays and skates. Even stranger is this question: is a shark a fish? The website of an influential European shark conservation organisation asks:

> Shark or Fish? Sharks are commonly termed fish, even though they are only distantly related to the classical (bony) fish. The evolutionary lines of the cartilaginous and bony fish separated about 400 million years ago.[6]

A pioneering twentieth-century American shark authority, Harold McCormick, wrote that, 'there are even leading ichthyologists who do not regard [sharks] as fishes at all but rather as representing a separate and distinct *Class* of life'.[7] Many sharks bear a superficial resemblance to dolphins, which are mammals; and the largest of all is the whale shark, not a very fish-like name. If a shark is not a fish, however, then what is it?

The origins of sharks, and their fantastically long journey into the twenty-first century—which so threatens them through destructive human activity—is a story that is in equal measure breathtaking and, now, tragic. And because it is a complex story it should begin simply, with some terminology.

Non-plant marine life is classified as either vertebrate or invertebrate, that is, without or with a backbone. As shown in the table on the next page, there are six classes of marine vertebrates.

Osteichthyes (ostee-ick-thez)	The bony fishes. Most of the 22 000 known species belong to one infraclass, the teleosts.
Chondrichthyes (kon-drick-thez)	The cartilaginous fishes. More than 1200 known species, grouped in two infraclasses: • elasmobranchs (sharks, skates, rays) • holocephali (chimaeras)
Agnathas	Two species of jawless fishes, the lamprey and the hagfish
Mammals	Approximately 120 species grouped in several infraclasses: • pinnipeds • cetaceans • sirenians • sea otters • polar bears, generally grouped with marine mammals
Reptiles	Approximately 45 known species grouped in several infraclasses: • sea turtles • iguanas • snakes • crocodilians
Birds	Many thousands of species, including cormorants, gulls, albatrosses, terns, shearwaters, guillemots, puffins and penguins

There are significant differences between elasmobranchs and teleosts. What they do have in common, apart from their shared liquid environment, is this broad definition of a fish:

'a poikilothermic [varying body temperature], aquatic chordate [backbone] with appendages (when present) developed as fins, whose chief respiratory organs are gills and whose body is usually covered with scales . . . or more simply, a fish is an aquatic vertebrate with gills and with limbs in the shape of fins'.[8]

What they don't have in common is the result of hundreds of millions of years of unrelated evolution:

Sharks mate, resulting in the production of relatively few large, hard-cased eggs or live young after a gestation period of at least nine months and up to two years.	A female teleost ejects many thousands of tiny eggs into the water which are fertilised when the male ejects his sperm among the eggs. Hatching periods range from less than a week to several months.
Sharks take up to 20 years to mature and become sexually active.	Many teleost species mature in less than two years.
A shark's skeleton consists mainly of cartilage.	A teleost's skeleton consists mainly of bone.
A shark's liver is large and oil-filled, to provide buoyancy.	A teleost has an air-filled swim bladder to provide buoyancy.
A shark's gills are uncovered.	A teleost's gills are covered by a moveable, bony plate, the operculum.
A shark's skin is covered with modified teeth called dermal denticles, which poke through the skin and are then 'non-living', that is, fixed in size and shape.	A teleost's skin is covered with living scales, which grow as part of the skin.

Many shark species are solitary.	Most teleost species shoal.
Most shark species have large mouths, containing dozens or hundreds of teeth, which are shed regularly individually or in sets and immediately replaced.	Most teleost species have proportionally smaller mouths and fewer teeth, which are anchored individually in the jaw and not replaced as a set.
A shark's upper jaw is not fused to its cranium.	
Sharks regulate the amount of salt in their bodies by retaining high concentrations of urea in their blood.	Teleosts excrete their urea.
Sharks total less than five per cent of all fish species and are much longer lived than teleosts.	

Furthermore, while there are tiny sharks and tiny teleosts, there are many shark species that grow far larger than any teleost. The largest teleost, the sunfish, at over four metres and over 2000 kilograms, is dwarfed by its chondrichthyes equivalent, the whale shark, growing to at least 12 metres and weighing 15 000 kilograms.

Why such major differences? Although elasmobranchs and teleosts have travelled along widely divergent evolutionary paths, both journeys took place in a common environment, our planet's oceans, seas, rivers and lakes. Oceanographer Meredith Grant Gross has observed that with 70.8 per cent of the planet submerged, 'ocean is a thin film of water on a nearly smooth globe, interrupted here and there by continents'.[9] Approximately 60 per cent of the northern hemisphere is submerged, and 80 per cent of the southern hemisphere. Earth's longest mountain range is not the Himalayas but the submerged Mid-Atlantic Ridge which extends from north-east of Greenland to Bouvet Island just above Antarctica. Likewise, Earth's highest mountain is not Mount Everest but Hawaii's undersea Mauna Kea.

The average ocean depth is 3800 metres, while the average land elevation is just 840 metres. The maximum ocean depth is recorded at over 11 000 metres, in the Mariana Trench southwest of the Pacific island of Guam; by comparison Mount Everest's peak is 8850 metres above sea level. The Atlantic is a 'shallow' ocean, the Pacific a very deep ocean.

Oceans have depth layers. The surface or epipelagic zone is defined as being to a depth of about 200 metres and comprises just two per cent of the total body of oceanic water. It is also called the 'mixed' layer, as it is subject to atmospheric influences such as rainfall, evaporation, temperature variation and wind. The pycnocline or mesopelagic zone (the original 'twilight zone') descends to about 1000 metres. This layer of denser, more stable water separates the volatile surface waters from the remaining 75 per cent of ocean water, the deep zone. This zone has its own layers: bathypelagic (to 4000 metres, also known as the midnight zone); abyssopelagic (to 6000 metres); hadalpelagic (ocean trenches).

How did this water originate? There are at least three theories. In the wet-accretion hypothesis, Earth developed from silicate rocks with water trapped inside them. Endless volcanic eruptions turned this water into vapour, creating a dynamic moisture-laden atmosphere. As the planet cooled, this vapour turned into rain which, over millions of years, filled up the gigantic basins that became the three oceans and their smaller seas.

The late-veneer theory looks back nearly five billion years, when a vast disc of gas, dust and ice began to form into what became our solar system. The proto-system's rotation and energy formed comets at its outer edge. Millions upon millions of those comets, composed of almost equal proportions of ice and solids, were subsequently drawn back into the inner solar system and bombarded Earth. The young planet's warmth melted the ice and gravity retained it as water. But this theory suffers from the fact that there is a chemical difference between the hydrogen in Earth water and cometary water.

A third theory hypothesises a single violent encounter between Earth and a vast, water-carrying embryonic planet, a bit like a water-filled balloon. Recent research suggests that no more than 50 and as little as fifteen per cent of our planet's water originated from space.[10] Less than three per cent of that water is fresh, mostly in the form of ice. Lakes, rivers, groundwater and atmospheric liquid make up just 0.5 per cent of Earth's water.

Seawater's saltiness is a result of rainfall weathering exposed rocks and mountains, and washing their life-giving minerals into the rivers which feed the oceans. The continual evaporation of pure water from the surface of the oceans ensures that their salinity level remains more or less constant. On average, one kilogram of seawater contains 34.7 grams of salt (about one heaped tablespoon), half of which is chlorine, and a third sodium. Four other elements found in seawater are sulphur, magnesium, calcium and potassium. Seawater also contains dissolved sediments from atmospheric dust, marine biological processes and seafloor volcanic activity.

Salinity and temperature are critical for the establishment of life forms, because they affect the rate at which chemical reactions take place. Furthermore, many of the ocean's salts, especially calcium salts, are used to build skeletons, and seawater's nitrates and phosphates are extracted by plants, the first link in the oceanic food chain. Photosynthesis enables plants to grow at the depth to which sunlight penetrates. Phytoplankton—single-celled drifting algae, mainly dinoflagellates and diatoms—feed on dissolved organic and chemical detritus. They in turn are consumed by the filter-feeding zooplankton—the copepods (tiny, shrimp-like crustaceans) and euphausiids (krill), collectively known as the 'insects of the sea' because of their abundance and their importance as a food source. A primary feature of zooplankton is their vertical migration, as they rise to the ocean's upper layers by night to feed, and descend into safer, dark water by day.

There are also large, predatory plankton such as the gelatinous jellyfish. Beyond plankton, the oceanic food web becomes bewilderingly diverse: worms, crustaceans (including thermal vent spider crabs), bivalves (such as mussels), echinoderms (sea cucumbers and starfish), nudibranchs, cnidarians (corals, anenomes), cephalopods (octopus, cuttlefish, squid), reptiles (turtles, snakes, the saltwater crocodile), teleosts, elasmobranchs, pinnipeds, cetaceans (such as whales), mammals and seabirds.

Where do the elasmobranchs fit in? Almost everywhere, through their evolved ability to feed off most other oceanic life forms. Evolutionary processes ensured the early development of biologically complex and sophisticated predators and scavengers, preventing the oceans from becoming overpopulated and unhealthy. The description that follows of the evolution of the elasmobranchs requires some reference to the Geologic Timeline, which is structured primarily according to fossil records and measured in millions of years (mya).

Precambrian Time
4500–540 mya

- Hadean Eon (4500–3800 mya). Earth formed from colliding planetoids. Unstable crust, very little free oxygen in the atmosphere, constant cosmic bombardments.
- Archaean Eon (3800–2500 mya). Stabilising biosphere and oceans, single-celled stromatolites become the earliest life forms.
- Proterozoic Eon (2500–540 mya). Oceans shaped by proto-continents. A single supercontinent, Rodinia, breaks up about 600 mya. Emergence of complex multicellular organisms.

Lower Paleozoic Era
540–407 mya

- Cambrian Period (540–505 mya). Explosion of life forms: earliest aquatic plants, earliest aquatic animals such as corals and molluscs.
- Ordovician Period (505–440 mya). The Cambrian–Ordovician transition is marked by the first mass extinction event.
- Silurian Period (440–407 mya). The second mass extinction event, the Ordovician–Silurian transition, gives rise to the evolution of jawless fishes and the first land-based plants. The first jawed fishes date from the end of the Silurian and also the first sharklike dermal denticle fossils.

Upper Paleozoic Era
407–245 mya

- Devonian Period (407–360 mya). Amphibians, insects, proto-forests. First shark teeth fossils in the

Lower Devonian Period. Emergence of freshwater and marine predators including cladoselache, the best-known Paleozoic shark. This period ends in the third mass extinction event, the Devonian–Carboniferous transition.

- Carboniferous Period (360–286 mya). Seed-bearing plants, ferns, earliest land reptiles. Major shark radiation, including emergence of the eugeneodont and hybodont sharks and the holocephali (the extant chimaeras).
- Permian Period (286–245 mya). The end of this period is marked by the fourth mass extinction event, the Permian–Triassic Transition.

Mesozoic Era
245–65 mya

- Triassic Period (245–210 mya). Reptiles diversify, first dinosaurs, earliest mammals. Emergence of the neoselachians, the modern sharks. The fifth mass extinction event, the Triassic–Jurassic transition.
- Jurassic Period (210–144 mya). Dinosaurs dominant, first birds.
- Cretaceous Period (144–65 mya). Earliest modern mammals, flowering plants, reptile domination. Continuation of the hybodonts. The sixth mass extinction event 65 mya terminates many land and aquatic life forms.

Cenozoic Era
65 mya–present

- Tertiary Period (65–1.8 mya). Mammals dominant, birds diversify, first apes, first upright hominids.

- Quaternary Period (1.8 mya–present). Humans become dominant. The seventh mass extinction event, the Holocene Extinction, largely human-induced, is generally considered by the international scientific community to be well advanced.

The planet's earliest life forms included the stromatolites—rocklike structures built up of microorganisms, especially blue-green algae—which can be seen in the shallow saline waters of Western Australia. By the end of the Precambrian, Earth's oceans hosted primitive marine invertebrates such as those found in South Australia's Flinders Ranges, 'the first appearance of large, architecturally complex organisms in Earth history'.[11]

The fact that life originated in water threw up its own set of complications. Compared to movement through air, swimming is difficult:

> When a body moves through a fluid a hydrodynamic force acts on it. A component of the force acts backwards along the direction of motion; it resists the progress of the body and is known as the drag . . . the force also has a component at right angles to the direction of motion. This is known as the lift . . . When a fish swims drag acts on its body and must be overcome by a forward propulsive force . . . When the body bends not only does the tail move to the side but the head moves a little to the side as well; the tail wags the fish.[12]

The Cambrian Period gave rise to the evolutionary development that would allow active rather than passive aquatic movement: a flexible internal vertebral column to support an animal's shape and to enable sinuous forward movement—swimming. The evolution of a skeletal structure meant that the future choice of internal building materials—bone, cartilage, enamel, dentine—would be influenced by an organism's surrounding environment.

The first swimming animals to emerge during the Cambrian Period included:

- conodonts, eel-like creatures with a stiffening notochord, that is, a rod providing backbone firmness, the muscles around which would have moved the notochord when they contracted: the origin of swimming. Their teeth structure, and terminal eyes, suggest that they may have been predators;
- the arandaspis, the oldest known vertebrate. Its fossil was first discovered in central Australia in 1959 and named after the area's Aranda Aboriginal tribe. Finless and armoured, it probably foraged on the seabed, hoovering in food and using its tail for locomotion;
- agnathas, the first true vertebrates, which appeared towards the end of the Cambrian Period. The name means 'no jaws'. They generally had stout bodies, paired fins and external protective bony shields around their heads and gills. Unlike the extant cartilaginous agnathas—the hagfishes and lampreys—most prehistoric agnathas had bony skeletons; and
- ostracoderms. Typical agnathas, many species were about 30 centimetres long although there are some fossil specimens exceeding a metre in length. The jawless mouth meant that the ostracoderm either hoovered in food or was a filter feeder—again, unlike the hagfishes and lampreys, which are parasitic suction feeders. Behind the bony head plates the body was covered with dermal scales.

The known shark fossil record begins in the form of scales and teeth. To have teeth, an animal must have jaws. Jaws evolved from the front pairs of gill arches over about 50 million years, and the first jawed fishes date from the Silurian Period. It is thought that the front gills initially became modified to move the jawless mouth up and down to pump water more effectively over the gills. Teeth evolved over the same period from skin scales—dermal denticles—migrating permanently into the 'new' mouth.

The Devonian Period, which followed the emergence of the first jawed species, has been called the 'age of the fishes'. Some of the first to emerge were:

- xenacanths. Freshwater predator–scavengers, with an upper jaw attached at front and back to the skull and therefore fairly inflexible;
- ctenacanths. More evolved predators with firmer keels and pectoral fins made flexible by cartilaginous supports, both of which improved pursuit swimming. Like modern sharks, males had claspers;
- placoderms. With heavy external bony plating over the head and thorax, and toothlike crushing structures in their simple jaws, the placoderms were the main predators of the Devonian Period, which has led some researchers to claim them as the planet's proto-sharks, although their skeleton was bony, not cartilaginous;
- acanthids. With two spined dorsal fins, they are often called 'spiny sharks', though as their skeletons have bony elements, it is more likely that they are the true ancestors of the teleosts rather than the sharks; and
- cladoselaches. So-named because of their multi-structured teeth, the cladoselaches first appeared about 370 mya.

> They had cartilaginous skeletons and are possibly the true ancestors of sharks. They grew to about two metres, had two dorsal fins with short, strong spines, large pectoral fins, a strong caudal fin and sharp teeth for grasping prey—all the ingredients of a fast-moving predator. Cladoselache fossils are rare, but a number of almost perfectly preserved specimens were found on the southern shore of Lake Erie in the United States.

If the Devonian Period was the 'age of fishes', the Carboniferous Period has been called the 'Golden Age' of sharks, as they radiated into a variety of experimental forms. This may be because fossil evidence indicates that during much of this period the seas were relatively 'empty', following the extinction event that ended the Devonian Period, the cause of which remains unknown. The demise of the once-dominant placoderms made available niches for new forms of predator, the eugeneodontida.

Among the evolutionary experiments of these early elasmobranchs were appendages whose use has not yet been explained and probably never will be. Ichthyologists struggle to describe their forms and explain their functions: one species had an 'enormous, flat-topped dorsal fin bristling with enlarged scales. Basically, it looked like a fish with a brush sticking out of its back . . .'[13] Another had an 'outrageous dorsal fin—the shape of an ironing board—that it seems was part of courtship display as it is found in the males only. The top of this fin was covered in rough, tooth-shaped scales . . . Was this supposed to mimic a huge mouth and make the creature appear more frightening?'[14] The teeth of some of these protosharks were even more baffling:

> They formed a whorl erupting from the back of a semi-circular 'conveyor belt arrangement', but the teeth did not fall away at the front as in modern sharks. Instead, they were rotated under the apex of the lower jaw, and then back up into a cavity under the jaw where they were stored in a tight spiral.[15]
>
> The teeth . . . were arranged in a single row, the bases of each tooth being greatly enlarged and joined together in a curved symphysical whorl . . . the entire whorl grew outwards as each new tooth was formed. This gave the shark a pair of saw-edged shears that projected from the mouth, each composed of a series of teeth and tooth-bases with the youngest teeth at the base and the oldest at the tips.[16]

As evolutionary experiments they represent the birth of elasmobranch diversity:

> Falcatus and Damocles are two separate genera of sharks, each with long spines protruding from the head that were directed forward. One [fossil] limestone slab of Falcatus actually has two sharks, one on top of the other. The shark on top of the slab, which is devoid of any spine, is actually biting the spine of the shark beneath it. This may be an indication of courtship behavior.[17]

And we have to guess at some other anatomical oddities:

> *Squatinactis* seems fairly convincing as a torpedo-style hunter . . . the torpediniforms are largely ambush predators who lurk on or in bottom sediments. When they encounter suitable prey, they 'jump' on their expanded dorsal fins, sucking the victim in under the mantle, which

> is then folded over the prospective lunch. Unlike modern rays, squatinactids presumably lacked electric cooking, which may well have dictated a somewhat different design. This hunting strategy makes sense out of what might otherwise seem a bizarre architecture. The anteriorly directed dorsals make little hydrodynamic sense for swimming. However, if the objective is to surge upward and forward over a slow-moving Paleozoic fish, in a single movement, *Squatinactis* is very well designed.[18]

The fourth mass extinction event ending the Permian Period wiped out 90 per cent of the planet's life forms. It may have been caused by a meteorite hit, or extensive volcanism, the latter possibly causing the oceans' oxygen levels to fall dramatically. Some marine species did survive, one group in particular becoming dominant in both saltwater and freshwater systems. These were the hybodonts, which were probably better able than placoderms to prey upon the rapidly evolving teleosts, although the fossil record is not certain. Nor is there certainty over their eventual demise:

> Among the vast array of modern type sharks that occur at the end of the Cretaceous, Hybodus was a carry over from the older Carboniferous. The hybodont sharks were one of the first modern types of sharks to appear in the fossil record. Some paleontologists believe that during the Jurassic, the hybodont line diverged into today's modern shark lineage. The hybodonts though displayed some primitive and different characteristics than the main body of sharks of that time. Hybodont sharks . . . went into a worldwide decline during the Paleocene [early Tertiary].[19]

One way through the uncertainty of the elasmobranch lineage is to consider not so much from which ancestors modern sharks evolved, but how form evolved. Thus cladoselaches had:

- a long anterior mouth extending from the front of the snout to the gill slits;
- a simple and relatively weak jaw joint; and
- an undeveloped vertebral column.

The later hybodonts had:

- an underslung mouth;
- a protrusible jaw;
- strengthened jaw joints; and
- pectoral and pelvic fin cartilage structures increasing their flexibility.

Hybodonts lacked:

- solid non-compressible vertebrae.

The earliest freshwater fossils of teleosts date back to just under 400 million years ago, whereas the marine teleost record is considerably younger, dating to the mid-Triassic Period some 230 million years ago. Although the original ancestors of teleosts may have been the acanthids, they also bear an uncertain relationship to the placoderms, held by some to be the sharks' ancestors!

Proof of the success of the evolving elasmobranch form—cartilage, variegated teeth in increasingly flexible jaws, powerful paired fins, claspers for internal fertilisation, a heterocercal tail fin—came with its ability to also radically diversify by flattening and adapting to a benthic lifestyle. Again, this was to take advantage of new prey, in this case on or near the oceans' floors, especially bivalves. The first batoid fossils date back about 200 million years (following the fifth mass extinction event, which eliminated about 20 per cent of marine families), and they continued to evolve into the Tertiary Period, when a few species of stingrays began colonising freshwater systems about 50 million years ago. And at about this time the largest sharks of all, the filter feeders, also began to evolve, to take advantage of an inexhaustible food source, plankton.

Earth's sixth major extinction event, 65 million years ago, famously brought an end to the planet's domination by land dinosaurs and sea reptiles. With the disappearance of these giants, mammals began to increase in size, diversify in form and spread rapidly. Some returned to the seas, those mammals becoming yet another shark food source. An outsized shark, *Carcharodon megalodon*, which can be roughly translated as 'megatooth', lived between 20 and one million years ago (or less) as a specialist whale predator. Megalodon's record is limited to rare fossilised vertebrae, and more plentiful teeth, measuring as much as 18 centimetres in length. Despite the scarcity of the fossil evidence, palaeontologists estimate that megalodon grew up to 15 metres long and weighed as much as 50 tonnes. (The largest great white sharks are about 6 metres long and weigh up to 2000 kilograms.)

The fossil record suggests that by the beginning of the Miocene Epoch, about 23 million years ago, megalodon had become dominant as a super-predator, but it died out during the Pliocene–Pleistocene extinction event 1.6 million years ago (when the first modern humans appeared). Some people, however, believe that 'megatooth' may not be extinct, while shark scientists continue to debate its lineage and genus, including its relationship to the extant great white shark, *Carcharodon carcharias*, which first appeared about

11 million years ago and evolved into a specialist predator upon sea mammals, mainly pinnipeds. (The Carcharodon lineage can be traced back to about 65 mya.)

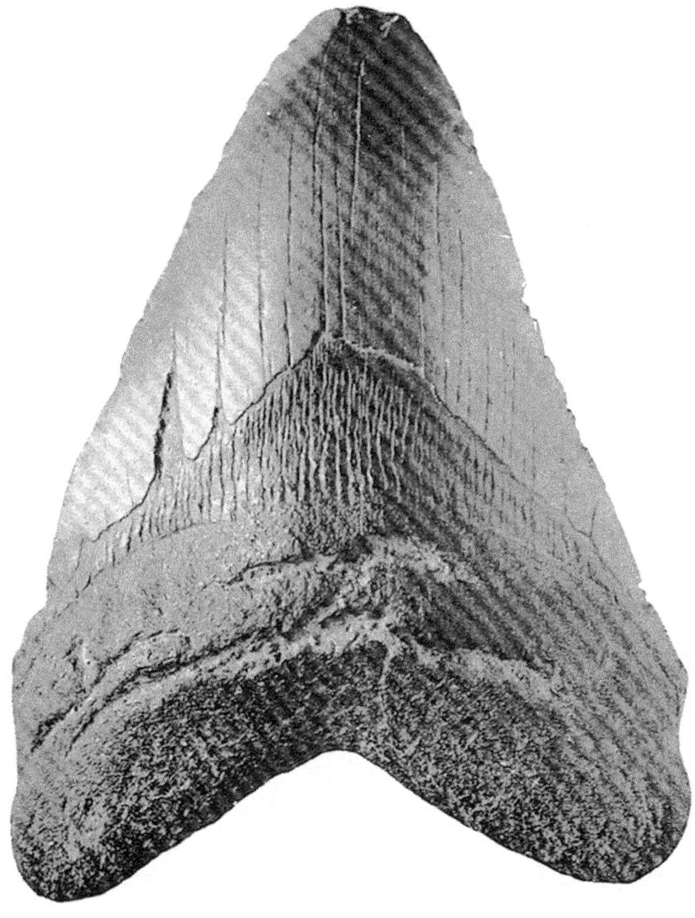

A tooth of Carcharodon megalodon, *the giant shark that grew to about 15 metres. This tooth is 12.5 centimetres long and ten centimetres at its widest.* (From *The Elasmobranch Fishes*, J. Frank Daniel, University of California Press, 1928)

Evidence for the continued existence of megalodon is slim. The well-known Australian ichthyologist David Stead wrote a lengthy account in his 1963 book, *Sharks and Rays of Australian Seas*, of a 1918 encounter between New South Wales fishermen and a gigantic shark. Stead linked the eyewitness accounts of these 'prosaic and rather stolid men, not given to "fish stories"', with his study of seemingly non-fossilised giant teeth dredged up from the Pacific Ocean and concluded that sharks of '80 to 90 feet long' might still inhabit the depths.[20] While there is still a degree of uncertainty as to what lives in those depths—as evidenced by the relatively recent discoveries of the coelacanth, the megamouth, and the colossal squid—it is possible that once megalodon's main prey, whales, evolved to inhabit cold as well as warm waters, this huge temperate-water shark died out from natural causes.

The taxonomy of fishes, cartilaginous and bony, remains a fluid and contentious science. New species are being discovered regularly and there is considerable scientific disagreement on many existing classifications. Such are the classification dynamics that a major reference work quoted in this book, *Sharks and Rays of Australia* (CSIRO, 1994), by P.R. Last and J.D. Stevens, has undergone significant updating in the relatively short period since its first publication; the second edition, published in 2009, formally names more than 100 new species (previously identified only by letters of the alphabet), and includes 26 new species discovered and named, major changes to the systematics of the dogfishes and skates, and significant new biological and species distribution information.

Presently the elasmobranchs are classed in fourteen orders, which are broken down into families; each family comprises one or more genus; each genus has one or more species (often very many species). The sharks can be further distinguished into two groups, the Galeomorphs (the first four orders in the table at the end of this chapter) and the Squaleomorphs, with the former considered to be more advanced in an evolutionary sense. The classifications in the table follow Leonard Compagno, the distinguished shark taxonomy expert, currently the Curator of Fish Collections, Iziko South African Museum, Cape Town.[21]

As a link between this chapter and the next, it is useful to be aware of the study of traditional or non-scientific classifications, known as ethnotaxonomy, or folk taxonomy. To take one example, a 2007 survey of Brazilian fishing communities found that cetaceans—whales and dolphins—were variously described as ‘fish’, ‘mammal’, ‘not-fish’, ‘fish-mammals’ or ‘like sharks’ family’, the latter owing to convergent evolution and the resultant physical similarities between dolphins and some sharks. Interestingly, ‘some fishers mentioned that despite having watched on TV programs the information that cetaceans are in fact mammals, they continue referring to whales and dolphins as fishes because they have learned it from the elders’.[22] Physical similarities aside, however, elasmobranchs are very different to both cetaceans and telecosts, as a closer look at their physiology and biology will show.

Group name	**Common name— Subclass Elasmobranchii**	**Order**	**Families**
Galeomorphs	Mackerel sharks	Lamniformes (*Lamna* = ancient Greek serpent-monster, *formes* = Latin 'shape')	Mackerel (mako, great white, porbeagle, salmon) Thresher Grey nurse (sand tiger/ Raggedtooth) Basking Goblin Megamouth Crocodile
	Ground sharks	Carcharhiniformes (Greek shark = *karcharias*)	Hammerhead Hound Barbeled hound Cat False cat Finback cat Weasel Requiem/whaler (blue, bull, tiger, reef)

	Bullhead sharks	Heterodontiformes (Heterodont = more than one type of teeth)	Single family (nine species)
	Carpet sharks	Orectolobiformes (Greek = long lobes)	Collared carpet Long-tailed carpet Blind Wobbegong Zebra Nurse Whale
Squaleomorphs	Dogfish sharks	Squaliformes (Latin *squalidus* = pale rough skin)	Dogfish Rough Gulper Sleeper Lantern Kitefin Bramble
	Sawsharks	Pristiophoriformes (Greek *pristis* = saw)	Single family (approximately nine species)

Group name	**Common name— Subclass Elasmobranchii**	**Order**	**Families**
	Angel sharks	Squatiniformes (Latin *squatina* = shark, skate)	Single family (approximately 19 species)
	Cow and frilled sharks	Hexanchiformes (Greek *exa* + *ankos* = six gills)	Frilled shark Cow (sixgill and sevengill sharks)
Batoids	Sawfishes	Pristiformes (Greek *pristis* = saw)	Single family (seven species)
	Wedgefishes	Rhiniformes (Greek *rhinos* = nose)	Single family (approximately seven species)
	Guitarfishes	Rhinobatiformes Greek *rhinos* = nose, *batis* = ray)	Three families (guitarfishes, thornbacks, panrays)
	Electric rays	Torpediniformes (from the Latin root for torpid)	Four families (numbfishes, sleeper rays, coffin rays, torpedo rays)

	Stingrays	Myliobatiformes	Nine families (stingarees, giant stingarees, sixgill stingrays, river stingrays, whiptail stingrays, butterfly rays, eagle rays, cownose rays, devil rays)
	Skates	Rajiformes (Latin *raja* = ray)	Three families (skates, softnose skates, legskates)
	Common name—Subclass Holocephali	**Order**	**Families**
	Chimaeras	Chimaeriformes (Latin = disparate body parts)	Ratfish Spookfish Elephant fish Rabbitfish Ghostshark

3
SHARK BIOLOGY

Form and Function

Most people—at least since the movie Jaws*—assume the creatures are solitary, stupid, antisocial brutes . . . Our research demonstrates that white sharks are intelligent, curious, oddly skittish creatures, whose social interactions and foraging behavior are more complex and sophisticated than anyone had imagined.*[1]

Cartilage and skeleton

The enduring and incorrect assumption that cartilaginous sharks are primitive, because of their great evolutionary age, also implies that the 'younger' teleosts, having bony skeletons, are more advanced life forms. The reality is that biological processes result in form-according-to-function diversity, and with fishes this is clearly evident in the endoskeleton. Furthermore, marine animals with bony skeletons—acanthids, placoderms—appeared in the fossil record before sharks arose.

The fossil record shows that from about 230 million years ago some marine fishes began to absorb and internally deposit hard calcium phosphate—bone—until eventually most fishes adapted to this form of skeleton (and ensured that land mammals have bony skeletons, having evolved from amphibious teleosts). Why did less than five per cent of the fishes not adapt to bone? One way of considering this is strictly in terms of function: evolution ensured that cartilaginous skeletons enabled mineralising calcium and dentine to be deposited elsewhere, specifically into mass production of predatory teeth. And perhaps there is an evolutionary parallel across the plains of Africa, where the vast majority of mammal species, herbivores, are preyed upon by a few carnivores.

So it is that biologically sharks are primarily distinguished from other fishes by having cartilage as their internal supporting architecture. Cartilage is ideal for pursuit predators requiring speed because it is light and flexible. And because cartilage does not provide a sufficiently rigid internal support base for muscular attachment, shark muscles also attach directly to the inner dermis, which means a larger area of support and, therefore, more muscle for speed and attack.

There are three fundamental differences between bone and cartilage. First, bone tissue is a combination of hard mineralised calcium, magnesium and phosphate (known collectively as hydroxyapatite) and flexible collagen fibres (collagen is the main protein in connective tissue). Cartilage contains collagen but lacks the mineralising elements. Second, unlike bone, cartilage has no blood vessels. Third, unlike bone, cartilage does not have blood-producing marrow cells. In the elasmobranchs, hematopoiesis (blood production) takes place in the spleen, in tissues around the gonads and, in some species, in Leydig's organ associated with the oesophagus.

Cartilage's basic building blocks are cells called chondrocytes, which produce and are eventually surrounded by an extracellular matrix of collagen, glycoproteins and complex carbohydrates. Water is also present in cartilage. Lacking a blood supply, chondrocytes are fed by diffusion. There are three types of cartilage, hyaline (gristle), fibrous (hard) and elastic (flexible). Elasmobranch cartilage is mostly hyaline.

The chondrocyte equivalents in bone formation are osteoblast cells which, because of their mineralising abilities, grow bone rather than cartilage. Despite these diferences the relationship between cartilage and bone is an intimate one. Juvenile mammal bones are mostly cartilage. Calcium in the mother's milk is deposited as calcium salts and the cartilage is gradually replaced by bone. In sharks, the vertebral centra become partially calcified by the calcium salts contained in seawater, as do the sharks' denticles, jaw, braincase, gill arches and fin supports. (Like tree trunks, calcified centra have annual growth rings, by which age can be determined.)

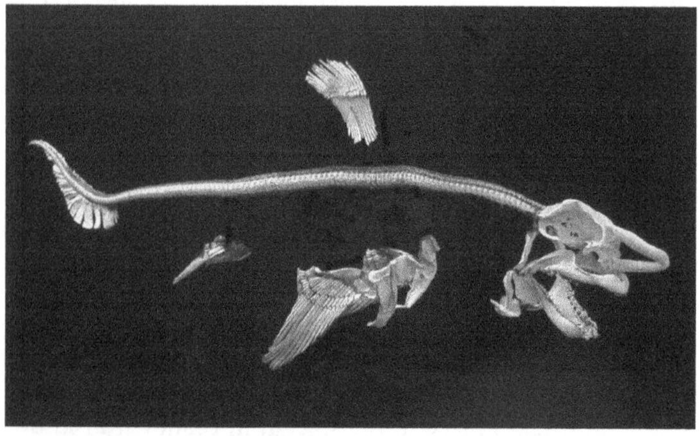

*The skeleton of a porbeagle shark (*Lamna nasus*).* (Dr Steven Campana, Bedford Institute of Oceanography, Canada)

The cartilaginous shark skeleton is an uncomplicated structure, comprising a braincase, jaw supports, gill supports, spinal column and various fin supports. The braincase has a base, side walls and a partial roof, a rostrum to support the snout and shaped parts for the eye and nasal sockets and the otic capsules. At the rear of the braincase the brain's highway, the spinal cord, exits through the *foramen magnum* ('big hole'). The paired gill arches which support the gill filaments—most sharks have five pairs—are positioned at the rear of the braincase.

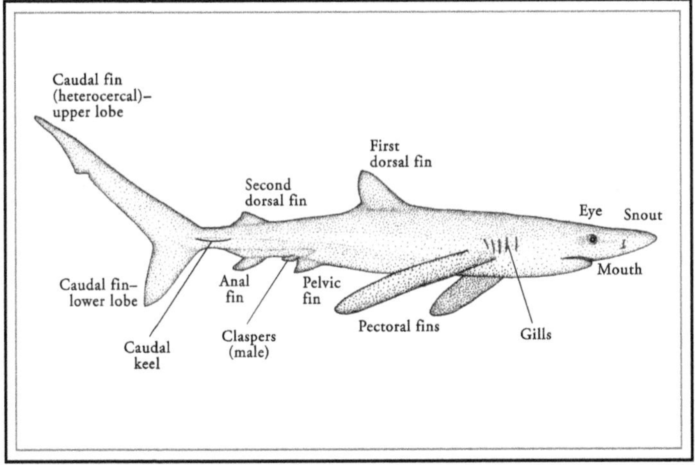

Parts of a shark

The spinal column consists of a flexible chain of linked vertebrae, more than 150 in some larger sharks. Each vertebra of the upper spinal column consists of a centrum through which the notochord runs, and a neural arch which protects the spinal cord. The arch is also designed so that muscles can attach to it. The caudal vertebrae bend upwards and so form the shark's tail. Each caudal vertebra also has a haemal arch, through which the dorsal aorta delivers oxygenated blood throughout the body. Fin supports in the form of girdles and radials attach to the spine. The u-shaped girdles support paired fins, while the radials are jointed and radiate out as spokes, in the shape of the fin. Over millions of years the proportion of the fin supported by radials had decreased significantly, allowing for greater flexibility.

Sensory processes

While there remains much to be learned about shark brains, tests have long since shown some species to have substantial learning capacities, in their responses to shape, sound, scent and colour. And tests have shown that sharks can differentiate quickly between human-set decoy prey and the real thing. Physically the mako, porbeagle, great white and hammerhead have proportionally larger brains than many of the so-called higher mammals, and these brains are required to process considerable amounts of information collected in remarkably sophisticated ways.

Sharks rely significantly on their sense of smell and ability to detect movement in the water around them, but many also have better eyesight than is commonly realised. Shark eyes are well adapted to water, a medium 80 times denser than air. They can see up to 25 metres in clear water and many species see in colour. Shark eyes have a visual streak across the retina that improves their vision of the underwater horizon. Not surprisingly, deepwater sharks tend to have big eyes, some with permanently dilated irises to allow maximum light penetration. Sharks' eyes, along with those of nocturnal vertabrates, have a feature called the tapetum lucidum. This reflecting layer behind the retina consists of thousands of minute platelets silvered by guanine crystals and enhances vision in dim light by bouncing the available light back to the retina, thus amplifying the image. In brighter light, dark pigments cover this layer so the eye doesn't take in too much light. Many shark species have a moveable eyelid, the nictitating membrane, to protect the eye from defensive actions of prey. Those which don't, including the great white, roll their eyes back in their sockets, exposing a hard pad at the back of the eye. This is why close-up photographs of feeding great whites often show them to have black, glistening, 'sightless' eyes.

Sharks also have a light-sensitive organ situated at the forefront of the brain, the epiphysis cerebri, more commonly known as the pineal gland or the mystical 'third eye'. The skin above this gland is translucent, allowing light to penetrate and stimulate the hormones. One theory is that this enhanced 'vision' allows sharks to process information about the seasons—because of their varying degrees of light—which is of fundamental importance for migration and mating.

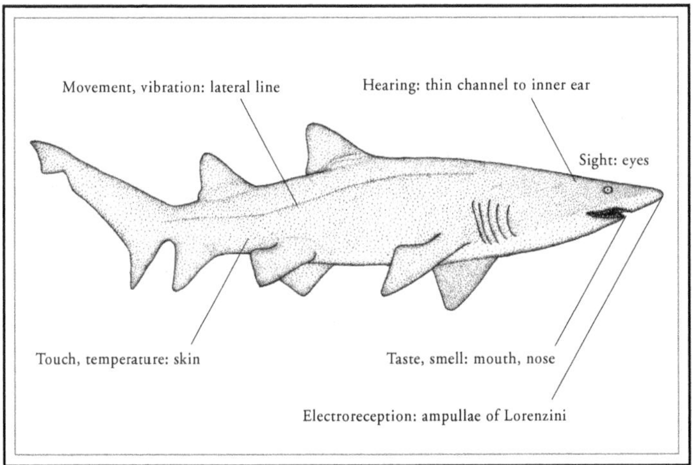

Seven senses of a shark

Sharks' nostrils are properly called 'nares' because they play no role in breathing. They are generally set forward on the snout, often with grooves leading from the mouth to facilitate water flow into them. The nares themselves have flaps which direct water flow into the olfactory lamellae—paired sacs packed with sensory hairs and nerve fibres that are acutely sensitive to chemical stimuli. Information about these stimuli is then transmitted through olfactory bulbs to the brain's large olfactory lobe. The size and complexity of this apparatus are clear indications of the importance of the sense of smell for the many species of shark which rely significantly on highly diluted chemical trails, both to seek prey and to communicate with their own kind.

Sharks do not have outer ears, but their inner ears play an important role in maintaining the animals' balance and equilibrium as well as detecting sound. Small openings behind the sharks' eyes allow seawater to flow through narrow tubes, the endolymphatic ducts, into the inner ears. The cavity of each inner ear contains three acutely sensitive fluid-filled cartilage tubes, and tiny calcium carbonate bones called otoliths, or ear stones. As the shark moves through the water, its brain picks up the signals of the fractional movement of these tubes and bones in their fluid-filled cavity. These signals tell the animal about its position in the water relative to gravity, as well as what its body is doing: accelerating or decelerating, pitching up or down, or moving from side to side. The otoliths also detect sound waves. Sharks are particularly receptive to irregular, low-frequency sound waves, such as those emitted by the erratic struggles of a fish or other creature in distress.

Sharks were once thought to have an otic system running the entire length of their bodies. While this has been proved not to be the case, they do have vibration-sensitive canals running along each flank just below the skin. This is the lateral line system. Open pores on the skin's surface allow seawater into tiny canals full of sensitive hairs which detect wave movements made by living things. The shark can then orientate itself in relation to the source of the wave movement. These canals also form complex patterns on the head.

Pit organs are minute sensory organs which bear similarities to the cells of the lateral line system. They are widely scattered inside pores or grooves across the upper body, including in species-specific patterns, which may hold a clue to their as-yet mysterious function. Another feature of pit organs is that they appear to be protected by distinctively shaped dermal denticles. Over the past century numerous theories have been ascribed to them, including external taste buds, detection of salinity variation, monitoring of swimming speed, monitoring of tidal currents, and prey detection.

Sharks, as predators, benefit from the fact that water is a good conductor of electricity. Even when a fish is motionless, or buried under sand, its muscles and heart emit electrical pulses. Clusters of visible pores on sharks' heads and snouts (which can bear a resemblance to a three-day growth) lead to jelly-filled canals which contain electroreceptors. These are called the ampullae of Lorenzini, after Stephan Lorenzini, who was an assistant to Nicolaus Steno. They not only detect minute electrical fields, but possibly also function as an internal compass tuned to Earth's electromagnetic fields.

A few species of rays create their own electricity, and a number of deep-water shark species are capable of bioluminescence, the chemical generation of light. These sensory abilities are used for predatory or defensive purposes, and will be addressed when describing the individual species, the electric rays (*Narcine* spp.), the cookie-cutter shark (*Isistius brasiliensis*) and the megamouth shark (*Megachasma pelagios*).

Movement

Elasmobranch skin, the integument, differs from that of teleosts in that it does not have scales. A shark's dermal denticles are individual growths rooted in the dermis, with enamel-tipped spines which break through the epidermis. Unlike scales, sharks' dermal denticles once grown are 'nonliving' and are regularly shed and replaced.[2] Their composition—dermal bone, dentine, enamel—classifies them as 'toothlike', as their name suggests, although they are more formally known as placoid scales. Dermal denticles grow in a profusion of shapes and sizes: they can be rounded (snout); shield-like (belly); keeled (flanks); diamond-shaped (leading edge of fins); or tiny and quadrangular (pharynx). Backward-facing, their primary purpose is to ease the animal's smooth passage through the water. Furthermore, 'possible and probable derivatives of dermal denticles include thorns, rostral sawteeth, fin spines, stings, clasper spines, and gill raker denticles'.[3]

The denticles' arrangement may also have evolved to render the predator 'hydrodynamically quiet'.[4] In some species, the denticles have an obvious defensive function, being sharp or abrasive enough to inflict considerable pain and wounds. Large pelagic predators tend to have fewer and smaller denticles, being required less for defence than to facilitate rapid movement through water. Most ray species, and a few shark species, have reduced numbers of denticles and their skin is instead often covered with a layer of mucus. This protects the slow-moving or frequently motionless animals from marine parasites.

Shark skin colouring and patterning are very varied, but

usually designed for camouflage, in both predators and prey. Thus, the fast-swimming pelagic predators sport a range of countershades of blue-grey and white (hard to see from above, hard to see from below), while rock and reef bottom-dwelling predators such as the wobbegong have blotchy camouflage-patterned skins and seaweed-like tendrils.

The skin is multi-functional:

> Attached directly beneath the tough skin are the red and white muscles which . . . transform the cartilaginous architecture of a shark into a fluid, graceful art. Perhaps in part because it is so difficult to move through water efficiently, sharks are very muscular animals. Something on the order of 85 per cent of a 'typical' shark's body weight is muscle, compared with about 35 to 45 per cent for humans.[5]

Red muscles, which require oxygen (aerobic), are used for cruising; white muscles, which do not require oxygen (anaerobic), are used to generate bursts of speed. A deepwater species such as the goblin shark, which doesn't rely on speed, has flabby muscles. A select group—thresher, great white, porbeagle, two species of mako and the salmon shark (*Lamna ditropis*)—benefit from an endothermic circulatory system in which the blood's heat does not leave the body through the gills but is retained through a counter-current heat exchange system, the rete mirabile. Thus these cold-blooded fishes are intermittently warm-blooded, their body temperature higher than that of the water surrounding them, which enables the muscles to generate more energy. This is a considerable asset to a pursuit predator in cool water.

Sharks have five types of fins, paired or unpaired, each of which has a highly specific function associated with locomotion. The first dorsal fin, so instantly evocative, acts like an upside-down boat keel, keeping the animal stable and upright in the water. It can be remarkably flexible. High-speed film of great white sharks demonstrates that the animals 'warp and buckle this fin at will' to control their movements.[6]

The paired pectoral fins are also mobile and provide uplift in the water to counterbalance the downward thrust resulting from tail propulsion. Critical for manoeuvrability, in fast-swimming pelagic species the leading edges of these fins are narrow, with convex tops and flattened undersides. This increases water flow speed over the fins. Open-ocean migratory species such as the blue shark have huge pectoral fins, which allow them to glide great distances along the ocean currents, expending minimal energy.

The paired pelvic and anal fins serve to adjust water flows around the shark's body and increase stability. Many shark species have a second, usually smaller dorsal fin which also contributes to the animal's stability. Some species have dorsal fin spines, often poisonous, to deter would-be predators. In male sharks the pelvic fins support the claspers, the reproductive organs.

The tail—the caudal fin, comprising evenly or unevenly sized upper and lower lobes—is the shark's main propulsion system. Shark tails vary tremendously in shape, size and function, from the great scythes of thresher sharks to the narrow whips of stingrays. Many sharks have heterocercal (nonlunate) caudal fins. This means that the caudal fin has unequal lobes and the vertebral column turns upward into the larger lobe. Some, such as mackerel and basking sharks, have equally lobed homocercal (lunate) fins. In skates and rays, the caudal fins are neither heterocercal nor homocercal, but either drastically reduced in size or absent altogether.

The skates and rays are the result of a logical but nonetheless remarkable evolutionary development. Their platelike shapes are not single discs but greatly modified pectoral fins adapted to life on the seafloor (with a few exceptions such as the pelagic manta ray). The expanse of the disc enables the fish to glide just above the substrate with minimal energy expenditure. Oscillatory propulsion is the term used when the pectoral fins are flapped up and down like the wings of a bird. This is how the giant eagle and manta rays 'fly' through the open ocean. Some skates have pelvic fins modified into leg-like appendages, which punt the animal into motion off the floor; it then glides.

Mouth region

While breathing and biting may seem to have little in common, shark evolution determined otherwise. As described in the previous chapter, jaws evolved as a modification of the front pair of gills. The first primitive jaws were crushing structures, which became much more sophisticated once lined with teeth.

Most sharks have five pairs of gills (a few species have six or seven) on either side of the pharynx. When water enters the moving animal's mouth, it flows into the pharynx and out through the gill slits. These slits are screened by rows of filaments, on the surfaces of which are tiny growths called primary and secondary lamellae, so structured as to present a maximum surface area. As the water passes through the gills, these lamellae absorb its oxygen into the shark's blood, at the same time releasing carbon dioxide from the blood.

Many shark species, particularly the bottom-dwelling rays and skates that are frequently motionless with their mouths on the sea floor, have two further gill-like openings behind the eyes.

These are the spiracles, which pump in water independently of the mouth. The oxygen so absorbed is transported directly to the eyes and brain. In fast-swimming sharks, which have more water continuously entering their mouths and passing through their gills, the spiracles are either reduced in size or absent. The process of taking in water while moving is known as ram ventilation. The taking in of water while motionless is known as respiratory pumping, and is something that fast-swimming pelagic species are unable to do. This is why they must keep moving in order to obtain oxygen. The heart is located directly beneath the gills, ensuring the most efficient passage of oxygenated and deoxygenated blood between it and the gills.

Physiologically the snout is not part of the mouth but, like the gills, has a function intimately associated with the mouth. There was a longstanding belief that because the mouth is underslung, a shark had to turn sideways or even upside down in order to grab prey. In fact, the fusiform sharks take their prey front-on. The jaws, being loosely attached to the skull, are protrusible; with the upper jaw 'free to slide along a groove in the cranium',[7] initiated by the cartilaginous snout flexing up like a drawbridge. For example, 'the lemon shark is an active hunter capable of rapid acceleration. When approaching its target at speed, it brakes with its pectoral fins, raises its snout, drops its lower jaw, protrudes its upper jaw and teeth, and then jabs forward several times to get a good grip'.[8] And with the porbeagle,

> . . . the opening of the mouth causes a contraction of the front tissues of the lower jaw which makes the first rows of teeth protrude to the exterior, turning the mouth into a kind of hooked trap . . . Next, the top jaw is lowered, bringing into operation the teeth designed to lacerate the prey.[9]

Not only does the speed of the initial jaw protrusion provide an advantage to the predator in striking its prey; the shark is then able to protrude its upper jaw repeatedly while chasing the prey or group feeding. Furthermore the complexity of the jaw movements are such as to 'reorient the teeth for more effective biting and [to] permit a variety of functional novelties (e.g. chiseling, gouging, excavation)'.[10]

The business end of a shark is its mouth, and the teeth in it—even more than a dorsal fin slicing water—are instrumental in feeding our negative attitudes towards sharks (one tragic result of which was the targeting of the harmless grey nurse shark in New South Wales waters, throughout much of the twentieth century, because of its ferocious-looking teeth; the species is now faced with local extinction). But, whether fearing sharks' teeth or not, it is hard not to admire such jaw-dropping masterpieces of nature. There are two ways of considering sharks' teeth. The first is by their physical form, the second by their manner of production.

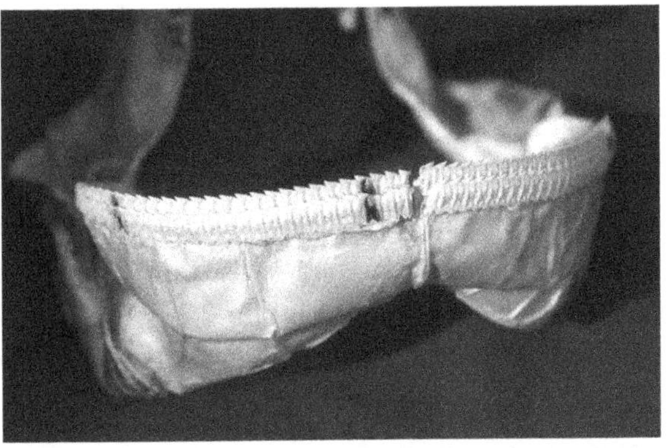

The jaws and teeth of a Greenland shark. In recent years research into this species has developed rapidly, through the work of the Canada-based Greenland Shark and Elasmobranch Education and Research Group. (Dr Chris Harvey-Clark, 2008)

Because sharks come in all shapes and sizes, and fill so many different niches in the marine biota, as a class they have developed an amazing variety of teeth types. There are four type categories: biters and shakers; crushers; grazers; a combination of any or all of these. Depending on their function they pierce, impale, seize, slice, clutch, clip, nibble and crush. They can be saw-edged, serrated, asymmetrical, broad, pavement-like, plated or lanceolate (lance-like). They can have prickles or knobs, be comb-like, awl-like, spine-like, thorn-like, cockscomb-shaped, inwardly curving, spindle-shaped or blade-like. And as if that wasn't enough, some are even 'bendable': the teeth of the bamboo shark (*Chiloseyllium* spp.) are normally erect and pointed for grabbing slippery prey such as squid, but can be folded flat in order to crush the shells of crabs, a major part of their diet, after which they spring erect again.

> Control of bending is built into the tooth, which has a thick root and a relatively small cusp sticking out. This configuration creates a lever that resists bending initially but flattens quickly once the tooth is pushed down hard enough.[11]

Being a predator, but lacking limbs and claws to help it seize and hold prey, a shark must have teeth that are always in prime condition. This is achieved by the teeth being continually shed and replaced. How is this done? First, shark teeth are not rooted in the jawbone but merely anchored to the skin by connective tissue. Second, while a shark has one or more functioning upright rows of teeth at any one time, replacement rows are developing in membranes on the inside of the jaw, deep in the mouth. As a row, or set, grows, it is pushed forward in the mouth by those developing behind it, until the upright functioning row (or rows) of teeth are ejected and replaced by brand new ones. Mechanically it is a conveyor belt arrangement. Replacement rates vary according to species. In some species new teeth appear as regularly as every eight to ten days. Some species of dogfish eject an entire set at once, while the cookie-cutter shark will sometimes swallow its entire set of teeth with a plug of flesh it has gouged from its prey. At any one time, functioning teeth in a shark's mouth number from a few dozen to many hundreds, which means that some species may shed 20 000 or more teeth over their lifetime. Of all their many attributes, that sharks throughout their lives have marching teeth in mobile jaws is surely cause for admiration.

It is speculated that shark teeth could even play a role in determining the edibility of potential food, since the greatest concentration of taste buds in a shark's mouth is in the lining right near the teeth. In other functions, a male shark uses its teeth to get a firm hold of the female's pectoral fin during mating; and territorial grey reef sharks are known to follow an agonistic display with a slashing teeth attack to drive off that threat.

Internal parts and digestive processes

A cartilaginous structure, the basihyal, is anchored to the floor of the mouth. It is erroneously likened to a tongue. In most species the basihyal is small and functionless, but some, such as the goblin shark (*Mitsukurina owstoni*), use it in conjunction with the muscles of the pharynx as a powerful prey-inhaling vacuum. The pharynx leads to the oesophagus which opens into the bag-like stomach. The pancreas secretes powerful digestive enzymes (mainly pepsin) and hydrochloric acid into the stomach to rapidly break down food. For reasons not fully understood sharks have an ability to retain food in their stomachs for long periods without breaking it down. Famously, the 1935 Shark Arm Case in New South Wales involved a large tiger shark which, two weeks after being caught and put on view in the Coogee Aquarium, spewed up a human arm with a piece of rope knotted to it, the arm having been sawn from the body. The arm was so well preserved that detectives were able to use a tattoo on the skin, of two boxers, to determine the identity of the murder victim.

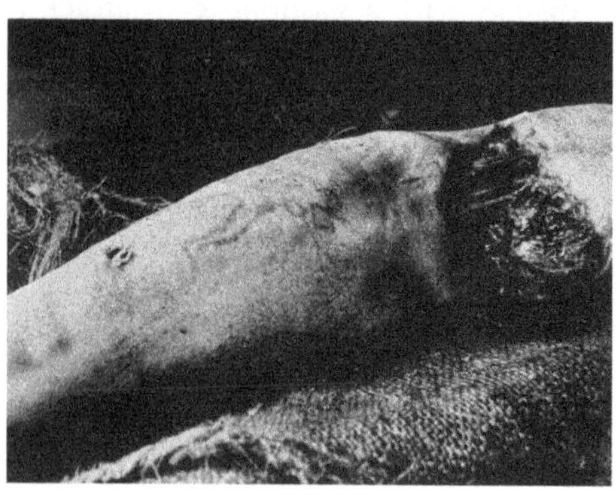

The tiger shark and the severed, tattooed arm of the famous 'Shark Arm Case'. (From *Shark Attack* by V.M. Coppleson, Angus & Robertson, 1958)

The stomach forms a double bend into the intestine, which is short and comprises spiral valves, their shape enabling greater surface contact with the food for the absorption of nutrients. Like many predators, sharks have symbiotic or parasitic relationships with a considerable variety of internal living organisms. Each of the shark's intestinal spirals may house a different species of tapeworm. Other parasites live in the stomach and feed off what the shark has consumed. Waste is excreted through the cloaca.

The ability of sharks to disgorge material in their stomach by everting the stomach bag is called voluntary stomach rinsing. The process is not often seen in action, as in this description:

> A lemon shark, *N. brevirostris*, was seen to evert its stomach naturally, where it hung limply out of one side of the mouth before head-shaking released an oily 'scum' from the surface of the stomach, which was then retracted slowly through the mouth and swallowed. The whole episode lasted 25–30 seconds . . . This process serves to remove parasites, indigestible material, toxic food, and . . . accumulated toxic metabolites.[12]

The largest internal shark organ is the liver, consisting of a left and a right lobe, and a much smaller median lobe, positioned beneath the oesophagus and stomach. Proportionately it is about twenty times larger than a human liver, and in some species can exceed one fifth of the total body weight of a shark. It is filled with squalene oil. Chemically squalene is an unsaturated hydrocarbon and also a triterpene, from which steroids derive. It is high in vitamins A and D. (The human body produces squalene as a skin lubricant.) It helps to fuel the shark's muscles and, because oil is lighter than water, provides the buoyancy that a shark would otherwise lack, having no swim bladder.

A requirement for living in water is that there is osmotic balance—equal concentrations of water and soluble substances—between the animal and its surrounding medium. Marine teleosts have far less salts in their bodies than seawater and therefore are constantly losing water through their skin and gills. To compensate, they replace this diffused body water by drinking almost constantly. The unwanted salt consumed in this way is excreted through salt glands near the gills. Elasmobranchs maintain the balance by a reverse biological process, involving their kidneys. Most of their nitrogenous waste is retained and converted into urea in the blood. Shark blood is on average about 2.5 per cent urea—almost 100 times that of teleosts. To prevent such high levels of urea damaging the shark's proteins, its blood also contains high levels of an organic compound, trimethylamine oxide, balancing out the urea and keeping the proteins stable.

Reproduction

Sharks reproduce by copulation. There is much to learn about this aspect of shark biology. Males have a pair of claspers, known as intromitent organs, between the pelvic fins. The male usually grasps the female's pectoral fin with its teeth in order to ensure that they remain together during copulation. Shark species mate in a variety of positions, including side-by-side, or with the male wound around the female's midriff. Male sperm passes from the testes along the Wolfian ducts to the seminal vesicle for storage. Siphon sacs near the claspers pump seawater into grooves in the claspers, the water transporting the semen into the female's cloaca and to her paired oviducts, her ovaries having released mature ova into them.

The gestation period after fertilisation is slow and varies between species, from about nine months to as long as 24 months. Birth numbers also vary considerably, from a single pup in some species to 300 in others. There are three methods of shark gestation and birth:

- Oviparous: after fertilisation a tough sac grows around the eggs before the female releases them through the cloaca. This egg sac is often known as a mermaid's purse and can be found in a variety of shapes, some of which have tendrils to help anchor the sac to the sea bed or rocks.
- Viviparous: the eggs hatch inside the mother and the young are nourished by a yolk-sac placenta until their live birth.
- Ovoviparous: the term traditionally used to describe the development of young inside eggs retained in the mother's body. They are nourished by the yolk sac. This term is being replaced in the literature with more specific descriptions relating to viviparity.

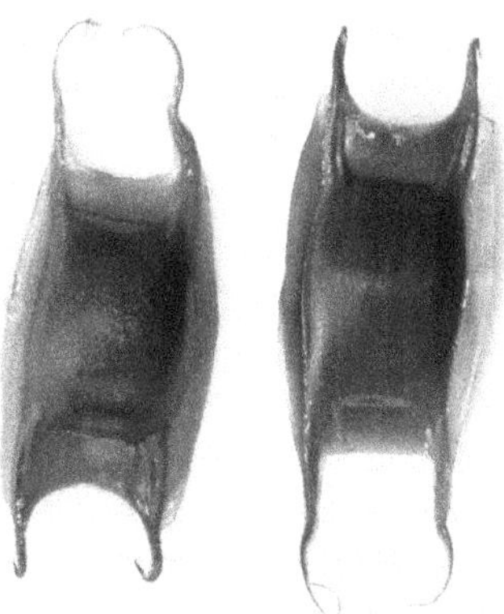

*Egg cases of the Antarctic starry skate (*Amblyraja georgiana*).* (M. Francis/NIWA)

Newborn sharks are immediately able to feed and there is no parental care. Indeed, the newborn can become prey to bigger sharks, including their own kind, which is why many species use pupping grounds that are either far from where adults of their species live, or in shallow waters that adults don't enter.

4

THE WAY OF THE SHARK ROADS

Sharks and Indigenous Societies

> *Now, at this time Moroa [creator] still held Lembe the shark between his thumb and forefinger at the base of the tail. Shark was tired of listening to Moroa's words but the creator had not finished giving his instructions. Moroa had not told Lembe the shark about the day-old bait fish which man would use to lure him to the side of the canoe. The shark did not want to listen so he jumped out of Moroa's hand and into the sea. Moroa was so angry at the shark who did not want to listen to his warning that he bent and picked up a handful of white sand and threw it at the shark. The sand stuck to his wet skin for the rest of time.*[1]

Dead sharks vanish. The unique physiology of sharks has left a fossil record that is patchy and inconclusive. This is because cartilage decomposes quickly. Unlike the teleosts, whose external scales and bony internal skeletons are major components of the marine fossil record, shark hard parts such as teeth, spines, calcified vertebral centra and small-to-tiny dermal denticles have had to tell much of the shark story. It is because of this that only recently has the economic—as opposed to the cultural—significance of elasmobranchs to indigenous societies become apparent.

Pre-industrial coastal societies across the Americas, Africa and the Indo-Pacific were familiar with sharks, skates and rays. They were generally reliable sources of food, while their skins, teeth and vertebrae could be put to a variety of practical and ornamental uses. Many indigenous societies also wove them into the social, cultural, spiritual and political fabrics of their lives. While those ancient traditions are not yet lost entirely, five centuries of European global exploration, conquest and economic and religious domination have either destroyed or greatly marginalised them. This chapter looks at the important role of sharks in indigenous societies from these three perspectives.

Food

It is tempting to speculate that the presence of sharks, rays and skates played a significant role in the establishment of coastal civilisations. The social groups who developed the first fixed pre-agricultural settlements chose sites where food, fresh water and shelter were readily obtainable. Favoured coastal locations were protected bays and estuaries—ideal elasmobranch habitats. Culturally complex societies began to develop at such sites, although archaeological studies of faunal remains found in coastal middens show little evidence of elasmobranchs, in comparison with teleosts, shellfish, mammals and birds. Given the skeletal nature of the elasmobranchs, however, this is not surprising. A seminal 2002 study carried out by Californian and Oregon researchers was the first to point out that academic publications 'focused on the methods of reconstructing the economic significance of elasmobranchs . . . are virtually non-existent'.[2]

From about the mid-1970s scientists began in earnest to try to quantify the importance of marine resources to the world's prehistoric societies. The primary analysis methods of midden sites include species identification, number of individual specimens present, raw weight of remains, and bone-to-meat ratios (meat yield from a particular species). Results showed a minimal presence of elasmobranchs but were, in all likelihood, a reflection of an unintentionally skewed methodology which was mostly reliant on skeletal remains.

> Only six types of [elasmobranch] elements—centra, teeth, dermal denticles, fin ray spines, tail spines, and rostral cartilage—are found in archaeological sites, and some of these are either too small to recover or are not present in all species of elasmobranch. In contrast . . . 26 elements from bony fish [are] commonly encountered in archaeological sites.[3]

Despite the scarcity of physical evidence, when the Californian and Oregon researchers began to consider the relative meat yields of the species they could identify, they found that elasmobranchs would have been an important target of prehistoric fisheries. Sharks, skates and rays have a comparatively high ratio of 'total weight and edible meat weight to the weight of their bones and teeth'.[4] A 'top-ten' (in descending order) Pacific coast table of species analysed reveals just four teleosts:

bat ray
spiny dogfish

brown smoothhound
shovelnose guitarfish
herring (teleost)
salmon (teleost)
thornback
California halibut (teleost)
angel shark
white croaker (teleost)[5]

It is therefore likely that prehistoric coastal societies would have fully appreciated the nutritional value of elasmobranchs. Indeed, in parts of coastal Australia, traditional Aboriginal clans continue to place a high value on stingrays, and some shark species, as food items.

Australia's indigenous peoples arrived in the continent as long as 75 000 years ago and have been described as 'the world's largest and most successful group of hunter–gatherers' and 'preagriculturalist[s] who had adapted extraordinarily well to life in a variety of habitats ranging from tropical forests, coastal and riverine environments, savannah woodlands, and grasslands to harsh, hot, and very arid deserts'.[6] A notable Australian indigenous seafood preparation combines the meat and liver of a stingray. The meat is first roasted or boiled, shredded when cooked and soaked in fresh water, then rinsed in seawater and squeezed dry. This removes any of the residual ammonia from the flesh. The nutritionally valuable liver, cooked or raw depending on local preference, is then kneaded into the shredded meat. The resulting mixture 'which tastes like succulent buttered crabmeat, is then separated into small round cakes and shared with relatives. This recipe is remarkably consistent throughout societies along the northern coastlines . . . Aboriginal groups separate the sharks and rays from bony fishes based on this unusual cooking technique'.[7]

Across the Tasman Sea, sharks once formed an important part of the diet of coastal Māori tribes of New Zealand. A mid-nineteenth-century account of shark fishing, as practised by the northern Te Rarawa tribe at Rangaunu Harbour, northern New Zealand, comes from the diaries of naturalist R.H. Matthews. Over just two nights in January each year (a strictly enforced rule), a fleet of about 50 canoes would take about 3500 dogfish sharks. Once all the sharks had been landed on the shore, the fins and tails were notched for owner identification and processing began:

> The *pane* (heads) were first removed; then a strip was cut following the curve of the belly. This strip, called the *whauaro*, was considered a great delicacy. The bodies were hung by the tails to a *tarawa* (a tall scaffolding), or thrown across a top rail, belly side up. There they remained until thoroughly dried by the sun and wind. The heads and *tapiki* (entrails) were generally left on the scene of operations. In a day or two the stench would be intolerable. The livers were thrown into a large funnel, made of green flax-leaves with a lining of soft fern-leaves, and suspended in a rough framework of tea-tree. Large stones were then heated and placed on them, and the oil was caught in calabashes. Surplus livers were put into the stomachs of the sharks, and hung up in the sun until the oil exuded from them . . . The dried sharks were stacked in food-houses, or *whatas*, just like so much firewood. Narrow strips were cut and cooked on hot stones, and beaten with a *paoi* (pestle for pounding fern-root) to soften the flesh. Sometimes the cooking was done in a

hangi, or steam-oven. In this case the flesh was cut in chunks, and not pounded.[8]

Inventive preparation techniques meant that almost all of the shark could be consumed. Drying, dehydrating, bleaching, salting, smoking and boiling rendered shark skin, stomach, cartilage, heart and fins edible. Some of these traditions continue—shark stomach is eaten in the Solomon Islands, Uruguay and some other countries, *nikigori* (gelatinous skin) is eaten in Japan, 'fish lips' (soft skin) in Singapore and Malaysia, and fin cartilage in Asian soups. (The controversial practice of finning, for soup, will be addressed in Chapter 7.) Gulper shark eggs are eaten in the Maldives, and salmon shark heart 'is prepared as *sashimi* in Kesennuma, Japan'.[9] Some societies considered that those who ate sharks' flesh would assume the animals' power and ferocity. Hawaiian chiefs reputedly ate the eyes of man-eating shark species, believing that this enabled them to foretell future events. And not surprisingly sharks were a particularly important food resource for island peoples, especially on the thousands of inhabited Pacific Ocean islands and atolls, many of which had limited natural land food resources such as edible plants or indigenous wildlife.

The most unusual traditional indigenous form of preparation, and one famous for being infamous, is the Icelandic *hákarl*, the origin of which is centuries old and may have come about by trial and error, as a way of rendering palatable the meat of the Greenland shark *(Somniosus microcephalus)*, a huge and common animal in a region of scarce edible resources. The flesh of many shark species has to be purged of its foul-tasting and smelling ammonia (from uric acid) before it can be eaten. The flesh of the Greenland shark is doubly toxic, in that it contains high concentrations of trimethylamine oxide, a protein stabiliser and anti-freeze. (An Inuit legend has it that long ago an old woman washed her hair in urine and dried it with a cloth which then floated out to sea and became the Greenland shark.) The traditional recipe for *hákarl*:

> Take one large shark, gut and discard the fins, tail, innards, the cartilage and the head . . . Cut flesh into large pieces. Wash in running water to get all slime and blood off. Dig a large hole in coarse gravel, preferably down by the sea and far from the nearest inhabited house—this is to make sure the smell doesn't bother anybody. Put in the shark pieces, and press them well together . . . Cover with more gravel and put heavy rocks on top to press down. Leave for 6–7 weeks (in summer) to 2–3 months (in winter). During this time, fluid will drain from the shark flesh, and putrefaction will set in. When the shark is soft and smells like ammonia, remove from the gravel, wash, and hang in a drying shack . . . Let it hang until it is firm and fairly dry: 2–4 months . . . Cured shark smells worse than it tastes. The texture is somewhat like a piece of fat, the colour is a dirty white/beige, and the taste reminds some people of strong cheese with a fishlike aftertaste . . . *hákarl* has been known to cause an involuntary gagging reaction . . . [10]

The patience required to render the Greenland shark edible was matched in the southern hemisphere by a degree of ingenuity required to catch large sharks. Oceania comprises three island groups: the northernmost Federated States of Micronesia (from Palau in the west to Kiribati in the east); Melanesia (including the Solomon Islands, New Caledonia, Vanuatu and Fiji);

and Polynesia, best described by its geographical descriptor the Polynesian Triangle, being New Zealand in the southwest, to Hawaii in the northeast, to Easter Island in the southeast.

Indigenous seafarers navigated this vast region of many languages and cultures for thousands of years, their paths crisscrossing, one result being similar methods of shark fishing. In 1927, the American Museum of Natural History published an exhaustive, 158-page paper by a noted ichthyologist, E.W. Gudger, which describes in detail nearly 100 types of wooden shark hooks used by the peoples of Oceania.[11] Gudger drew on many sources, including museum collections in Australia, Hawaii and the continental United States: 'In addition to all these collections of hitherto undescribed material with which I have had the pleasure of working, the literature has been thoroughly researched and all known and many hitherto unknown descriptions and figures have been brought to light and incorporated herein'.[12] He also described hooks collected by 'Salem sea captains during more than one hundred years past in their rovings through the South Seas'.[13]

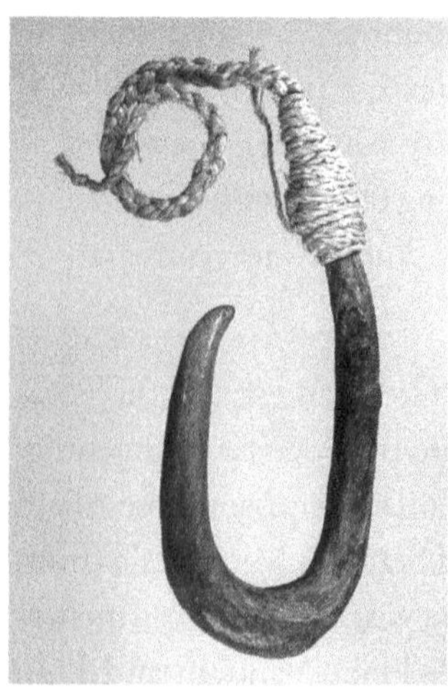

A hardwood and cord shark hook, collected from either Tonga or the Society Islands during one of James Cook's Pacific voyages undertaken between 1768 and 1780. The unevenness of the shank, showing marks from off-shoots, indicates that it is made out of a rhizome, typical of a hook used to catch sharks. (Georg-August University, Göttingen, reproduced by kind permission of Dr Gundolf Krüger)

These wooden hooks, in societies lacking metal, were manufactured with the same precision and wisdom lavished on the best modern trout flies. As well as sharks, wooden hooks were used to catch the oilfish (*Ruvettus pretiosus*), a large deep-sea teleost with quality flesh, whose liver is still valued for the purgative qualities of its oil. The shank leg and barb leg of these hooks were either carved from a V-shaped fork of a tree limb or, ingeniously, grown from saplings of trees such as mangroves and casuarinas:

> The *fe* or shark hook was made from a shrub, the *tiere*, which when it reached the height of about three feet, was twisted into an open knot, with a diameter of about 5 inches; it was then allowed to grow for about two years before being cut. The hook was then shaped, and a piece of hard wood spliced on as a barb projecting inwards. The bait was tied on over the barb; the fish working at this, as the wood was springy, gradually got its jaw between the barb and the stem of the hook. On being struck the barb caught in the gills, and the fish was hauled up sideways.[14]

The shark hook barbs were made from elements such as bone and teeth. The snood, or cord of attachment, was frequently made of coconut sennit—braided strands of fibre. Some of these hooks were huge: Gudger records one at Milne Bay, New Guinea, which was nineteen inches long.[15] The Hawaiian *makau mano*, 'used by their old kings to catch sharks . . . was not infrequently baited with human flesh for shark fishing, a slave being sacrificed for the purpose'.[16] Stone sinkers were attached to catch bottom feeders. The practical importance of these artefacts, and their value to their owners, is evident in this 1888 observation: 'Fishing, moreover, seems to be one of the principal industries of Trobriand, shark fishing in particular being practiced. This is shown by the large foot and a half long wooden hooks . . . which the natives are extremely reluctant to part with'.[17]

Other shark fishing methods practised across the Pacific region were spearing, trapping, hand catching, darting, netting and poisoning.[18] The most amazing, though, was shark calling: the art of luring a shark to a canoe by use of sound and then coaxing it into a snare. There can be few more artful methods of harvesting creatures from the wild and, not surprisingly, the practice had considerable religious and cultural significance. In the early 1970s, Australian-born artist and author Glenys Köhnke spent time with an elderly man known only as the 'Old One', who lived in a remote village on the west coast of the Papua New Guinean island New Ireland, and was possibly the world's last shark caller. She described the process of 'the one who knows the way of the shark':

> The shark caller's equipment is simple but very effective. He has a dugout canoe with outrigger for stability and a wooden float carved from light kapiak wood to resemble a two bladed propeller. This has a hole burnt through the centre part through which a plaited cane rope passes to form a noose on the underside and a handle on the top side. This float called kasaman and noose is used to snare the shark. He has a cane lure stick, lenantulus, to which he attaches the lure fish. The lure stick is short and light and easy to manage with one hand. He also takes larung, the rattle made from half coconut shells threaded onto a cane hoop . . . the sharks are attracted to the side of the canoe with the coconut rattle and lured into an open-plaited cane noose. The noose is attached to the wooden

propeller-shaped float. The shark caller holds the float above the surface with the noose underwater. The man passes a lure fish on a pole through the noose and offers it to the approaching shark. As the shark advances, the man draws the fish through the noose and the shark follows. When the shark is through the noose up to his pectoral fins, the shark caller drops the float onto the shark's back and tightens the noose by jerking the cane handle on the top side of the float, upwards. Then he casts the shark away to fight the propeller float. The float does not spin, but offers a great deal of resistance to the shark which tries to dive or scrape it off along the surface of the water. This tightens the noose around the shark. When it is exhausted the light kapiak wood float will bring the shark to the surface. The shark caller will paddle up to it. With some species of shark he first spears the eyes, but the actual killing is done with a solid wooden club.[19]

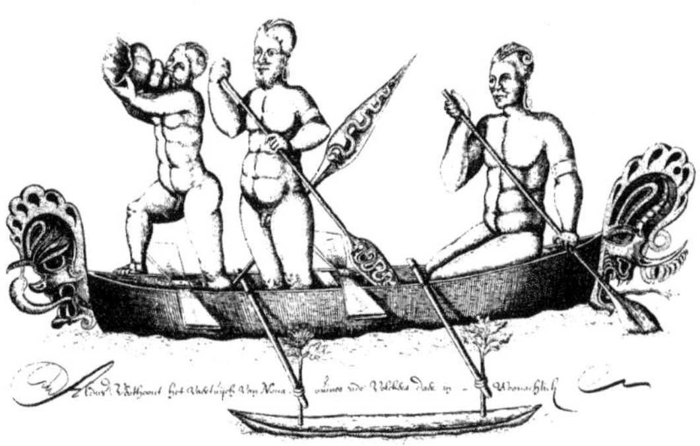

This illustration in Abel Janszoon Tasman's Journal, *dated April 1643, depicts a scene off the west coast of New Ireland. It is captioned, 'A view of a vessel of Noua Guinea, with the natives living there.'* Note the propellor float across the middle of the boat. (From *The Shark Callers*, Glenys Köhnke, Yumi Press, 1974)

Furthermore, the shark callers' intimate knowledge of sharks and the sea allowed them to make use of shark roads. In the words of the Old One:

> There are shark roads in the ocean in the season of lamat [calm season of the dry reef] when men go to call the sharks. The roads are broad, calm tracks of water in which man may see his face looking back up at him. It is only on these special roads that the sound of larung [coconut rattle] will carry great distances . . . When larung is shaken just under the surface of the water and knocked slightly against the side of the canoe it sends out many tiny waves which travel like the rays of the sun, out along the smooth surface of the shark roads. The song which larung makes travels far along the roads to where the shark is swimming. He will come to investigate the call of larung. No one knows what the shark thinks larung is. Ah, he may think it is the rush of tiny fish jumping out of the path of a larger fish. He may think it is the diving of the tarangau [sea bird] into the water to feast on a school of fish . . . There is magic too . . . Without the magic, larung could call all day and the shark would not come to it.[20]

Teeth, tools, saws and other practical applications

Indigenous societies valued sharks and rays for much more than their meat. At least eight distinct body parts had separate, practical functions in traditional societies. Shark teeth come in every conceivable shape and size, so can be put to a great variety of uses. Single teeth set into handles became tools for tattooists and surgeons, for piercing, drilling, sawing and carving. Shark teeth dating from the earliest civilisations have been found hundreds of kilometres from their place of origin. Teeth sourced to Red Sea sharks indicate early trade links between the peoples of the Nile and Mesopotamia. In prehistoric south Florida, shark teeth were used for cutting, carving, hafting and engraving in place of the hard metals and workable stone that were scarce in the area. The coastal Calusa Indians traded them inland, particularly bull and tiger shark teeth.

Shark teeth set into spears or clubs made formidable weapons, as did shark-tooth-studded gloves. In the 1830s, William Ellis, an American missionary, recorded Tahitians using a small cane studded with sharks' teeth to flagellate themselves in times of grief. He also described a short fighting sword: 'instead of one blade it had three, four or five. It was usually made of a forked *aito* branch; the central and exterior branches, after having been pointed and polished, were armed along the outside with a thick line of sharks' teeth, very firmly fixed in the wood'.[21]

One elasmobranch, the sawfish (*Pristis* spp.), has a ready-made sword. (There is no known cultural use by pre-industrial societies of the much smaller sawshark (*Pristiophorus* spp.) rostrum, presumably because its deepwater benthic habitat put it out of reach of fishers.) The sawfish rostrum contains sets of evenly spaced, long, narrow teeth. Archaeological studies of midden sites along the coast of Brazil indicate that local hunter–gatherer tribes used these rostral teeth as arrow tips and harpoons. They modified the teeth for a particular purpose, either by drilling holes in them, abrading their surfaces, or filing or cutting their roots to get sharper and more slender instruments.[22] A large sawfish rostrum can exceed 1.5 metres in length—a formidable weapon. Northern Australian Aborigines used them as fighting clubs. Another favoured weapon was the long, sharp and poisonous ray spine. In Australia's Top End, spear hafts were tipped 'with a bristling bouquet of venomous stingray spines. Wounds caused by this fearsome weapon were nearly always fatal'.[23] And Cape York's Wik peoples 'would sometimes cut rings from the tails of thorny rays, creating spiny "brass knuckles" which made punches more dangerous during fighting'.[24]

British-born Walter Roth, a physician and anthropologist who was appointed the protector of Aborigines in Queensland in the late nineteenth century, travelled widely in northern Queensland recording Aboriginal cultures. He also collected many artefacts, including shark-teeth knives, now in the Australian Museum, that he took from the Gulf of Carpentaria coastline between the Mitchell and Staaten Rivers. These knives

> . . . were made of an elongate piece of ironwood with a slot in one side where eight to nine shark teeth were inserted and fixed with adhesive. Adhesive also was found on the rounded and grip ends of the knife. It was at the grip end that looped handspun bark fibre string was wound round and attached with adhesive. Roth said that when a man used this weapon, he first hid it from view, either in his left armpit, or hung it by a loop over his forehead so that it hung behind his neck and out of sight of his opponent. At close quarters the knife was brought out, and hacked into the victim's flank or buttocks. Roth reported seeing some of these weapons up to 20–23 cm long. He first saw one on the Palmer River, where it had been obtained from a man Roth identified as being a Kundara man living around the mouth of the Mitchell River. He said the man called it a *Kulkong* which he took to mean tooth. The knife was only used for hacking purposes, never for

sawing meat. Roth said the Gunanni people called it *kap-patora*. Roth noted that P. B. King had reported a similar weapon being used at King George Sound, Western Australia, in 1839.[25]

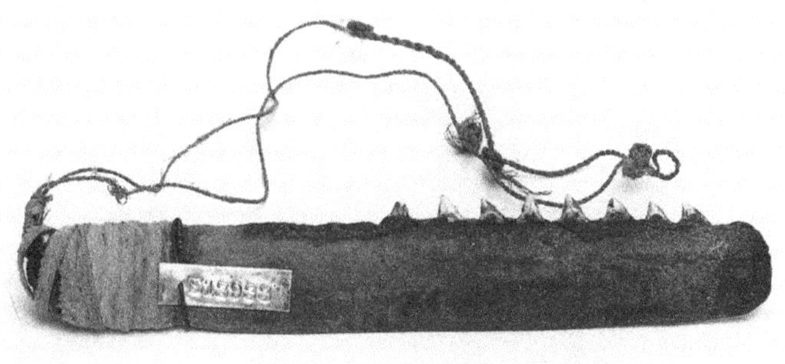

This shark-teeth knife is 20 × 3.7cm. Its formal description notes that 'a piece of white European cloth' has been wound round the knife. (Courtesy of the Australian Museum, citation [AMS391:M4080])

But shark products weren't only used for violent ends. The Anindilyakwa people of the island of Groote Eylandt in the Gulf of Carpentaria, west of Cape York, used small sawfish rostrums as hair combs. The calcified centra of the vertebral column made charming necklaces the world over. The beneficial effects of shark liver oil, rich in vitamin A, were recognised, and the oil was also used as a cosmetic. Some societies considered the claspers, the male shark's sexual organs, to have aphrodisiac qualities.

Curing shark and ray skin is a fairly basic drying process which results in a leather-like product. Most shark skins are studded with dermal denticles, and cured skin functioned effectively as sandpaper, although the denticles themselves could be sanded away to give a smoother finish. Cured sharkskin, although less flexible than mammal leather, could be shaped into rudimentary footwear. The Hawaiian *pahu* drum and the Sumatran tambourine were made of ray skin.[26] The Inuit wove durable ropes from the skin of the Greenland shark.

Shark teeth were not only used as weapons. According to the official description accompanying this image, shark teeth were so valuable that Māori traded them throughout the country. Furthermore, sharks were lassoed by the tail to avoid damaging the teeth, which were used to make necklaces. These are great white shark teeth. (Museum of New Zealand Te Papa Tongarewa, negative number ME015858)

Some of the teeth that were traded or bartered were used as tools rather than weapons or adornments. Bull and tiger shark teeth set into wooden boards formed prehistoric graters used for processing staple foods such as manioc and other tubers.[27] Charles Darwin recorded a similarly innovative use of the skin, in Tahiti. Darwin and members of the crew had enjoyed a convivial meal with their hosts, after which

> . . . a curious snuff was observed by Mr. Stokes, and from the method of using or taking it, I am inclined to think it an old custom, not imported by the white men. A substance, not unlike rhubarb in its appearance, but of a very pleasant fragrance, was rubbed on a piece of shark's skin, stretched on wood; and much it appeared to please an old man, who valued this snuff-stick so highly, that he would not part with it.[28]

In Japan a traditional and still preferred wasabi grater is made from shark skin stretched over a small wooden paddle—it is called *samekawa oroshi ki*, which translates as 'shark-skin grater tool'.

Spiritual, social and political applications

Australian Indigenous beliefs state that across the Australian continent in the period before creation, primal spirits interacted with one another to form and shape the land, sea and sky. Once their work was done they themselves became landscape features or animals and they handed down their sacred knowledge to mortal people, to nourish them and provide them with laws and norms. These creation beliefs are at once spiritual, cultural, social and political, and their legends explain the work of the creator beings. Across the vast continent, clans became associated with specific creator ancestors, and shark and ray totems are common among coastal Aboriginal clans. A number of the clans of the Yolngu people have as their totemic link to the creation time the whaler (bull) shark known as Mäṉa:

> According to the public version of the story, this ancestral being began his journey along the coast of northeast Arnhem Land. While sleeping on the beach, Mäṉa

> was speared by an ancestor from another clan who did not want other creator beings near him. Enraged by this stealthy attack, Mäṉa charged inland from the sea, exploding into the landscape. The ancestral shark gouged his way inland using his teeth to carve out several river systems. As he journeyed onward, his teeth broke off on the hard riverbanks; these lost teeth became the pandanus tree which line rivers today. The leaves of these trees are dagger-shaped with serrated edges, like shark teeth. These trees represent both Mäṉa's anger at being speared and the stingray-spine tipped spear that Mäṉa carried to avenge his death . . . This ancestral event explains why modern whaler sharks are dangerous and why some sharks still enter freshwater . . . The travels of this shark ancestor carried him through the lands of several related Yolngu clans. If these clans are on good terms politically, they may acknowledge that a single shark traveled through all the lands during a single journey, linking them. Alternately, when the clans wish to express their distinctive identity, they will explain that each clan's shark was a different ancestor, each starting its separate journey from the spearing incident.[29]

Thousands of kilometres to the east, across the Pacific Ocean, the Aztecs, a tribe of North American origin, had created a great civilisation in the Valley of Mexico, the heart of which was its capital city Tenochtitlán, the 'place of the prickly pear':

> This great city of Tenochtitlán is built on the salt lake, and no matter by what road you travel there are two leagues from the main body of the city to the mainland . . . The city itself is as big as Seville or Córdoba. The main streets are very wide and very straight; some of these are on the

> land, but the rest and all the smaller ones are half on land, half canals where they paddle their canoes. All the streets have openings in places so that the water may pass from one canal to another. Over all these openings, and some of them are very wide, there are bridges . . . There are, in all districts of this great city, many temples or houses for their idols. They are all very beautiful buildings . . . Amongst these temples there is one, the principal one, whose great size and magnificence no human tongue could describe, for it is so large that within the precincts, which are surrounded by very high walls, a town of some five hundred inhabitants could easily be built. All around inside this wall there are very elegant quarters with very large rooms and corridors where their priests live. There are as many as forty towers, all of which are so high that in the case of the largest there are fifty steps leading up to the main part of it and the most important of these towers is higher than that of the cathedral of Seville . . . [30]

The hidden irony in this enthusiastic description is that it was written by Hernando Cortès, soon to lead his Spanish *conquistadores* to bloody victory over the Aztecs, destroying Tenochtitlán in the process. Mexico City rose gradually over its ruins. In the late 1970s, a large and sophisticated pyramid—the Aztec Great Temple—was discovered beneath the very centre of the modern city. Among the many artefacts were remains of aquatic predators—sawfishes, sharks and crocodiles—and it is likely that these were a core feature of religious ceremonies. An Aztec creation myth held that the world was born through the violent tearing in half of a titan called Cipactlí. Humans lived on top of her floating lower half and her upper half was the heavens. Cipactlí's upper half required regular feeding in the form of human sacrifice, a perfect instrument for which was a large sawfish rostrum:

> In Aztec religion, they [rostrums] were powerful symbols representing the connection between the fecundity of the landscape and warfare . . . The use of sawfish snouts in ritual is detailed in a text written soon after the conquest of the Aztecs . . . In certain heart extraction sacrifices, the neck of the victim was crushed with the snout of a sawfish, preventing any inauspicious cries. Presumably, this action also allowed Cipactli to symbolically 'bite' the offering before the heart and blood were offered to the sun . . . these 'swords' of Cipactli were potent symbols of the Aztecs' obligation to fertilize the predatory, devouring earth with blood and bodies, so that she could in turn nourish mankind.[31]

Sharks and sawfish have less violent roles in other creation myths, as illustrated by this Tongan folktale:

> There is this shark-spirit who lives in the sea around Niuatoputapu who is named Seketoa. The descendants of Maatu, the chief of Niuatoputapu have the right to call on Seketoa and Seketoa will help them. When Maatu wants to speak with Seketoa he sends out his matapules (assistants) and they throw some kava root into the sea. Then two remoras will come to the kava roots. (Remoras are fish that live and will even ride on sharks.) These two remoras are the matapules of Seketoa. After the two remoras come they will go away, then a small shark comes and goes away. Then a larger shark comes and goes away. Finally a great big shark comes. This is Seketoa. Then Maatu, the chief, will speak with Seketoa.

> One night some Samoan ghosts stole the mountain from the island of Niuafo'ou. There is now a lake in the middle of Niuafo'ou where that mountain was at. These ghosts were taking the mountain by pulling it to Samoa. When they passed Niuatoputapu, Seketoa saw what they were doing. He sent his matapules to go near the ghosts and crow like roosters. Seketoa's plan was to fool the ghosts so that they would think it was almost morning. Then they would leave the mountain where it was at and hurry home to Samoa. When the Samoan ghosts heard the matapules of Seketoa crowing they said to themselves, 'Hurry, it is almost morning' and then they started to pull the mountain even harder.
>
> When Seketoa saw that his trick was not working, he swam to where the ghosts were pulling the mountain. He showed them his anus (mata tuungaiku) which was red. When the ghosts saw it, they thought that it was the sun coming up, so they put down the mountain where it was at and hurried back to Samoa. When they discovered that they had been tricked by Seketoa, they were so embarrassed that they never returned. The mountain that they left in the sea has now become the island of Tafahi which is several miles north of Niuatoputapu.[32]

Many Pacific Ocean island cultures recognised sharks as potent spiritual forces. In prehistoric Hawaii the importance of fish as the staple diet (edible land resources being limited to native bats and birds) elevated fishing to a sacred art. The capture of a shark, an intelligent, inquisitive creature of admirable strength, size and occasional ferocity, became the highest expression of that art. There were many Hawaiian shark gods, one of the better known being *Nanaue*—the offspring of a human woman and a shark king who could take human form. *Nanaue* had a shark's mouth between his shoulderblades, and an insatiable appetite for human flesh. (The legend of *Nanaue* perhaps bore some reference to cannibalism, or was a warning of the dangers of unnatural procreation.) A less voracious Hawaiian shark god, *Kane'ae*, took on human form to satisfy his love of dancing.

In what might be described as the Hawaiian equivalent of the Australian creation time, *'aumakua* are the benevolent spirits protecting individuals and families:

> *'Aumakua* belonged to and protected families, or a group of kinsmen, and passed from generation to generation. They were thought to be ancestors of these kinship groups . . . The early Hawaiians regarded certain sea animals, such as turtles, eels, squids, porpoises, and most notably sharks, as the physical embodiments of personal gods . . . If a species of shark were *'aumakua*, any of its members received offerings for special favors, such as good luck at sea and protection from drowning, prior to embarkation of a fishing expedition. Many fishermen, however, regularly fed a shark at a special spot along the shore or from a canoe and came to recognize them as individuals and even as pets.[33]

Such intimacy with the natural world—with sharks in particular—may seem hard to believe. Hawaiian author and publisher John Dominis Holt was born in 1920. As a young boy, he experienced sharks up close with a local elder:

> Off we went to the place in the reef where the sides slanted sharply to the bottom. Here we had to dive much deeper than before. Kai'a was old but extremely strong. He had

> dived all of his life and knew how to get down to the bottom quickly, with little exertion. I clung tightly to my old friend and *kahu*, and we passed through layers of sunlight in the water . . . It was darker down there, with shafts of light slipping through the cracks in the coral above and illuminating the sand in a dim glow. Then I saw these great living things lying on the bottom, rolling slowly from side to side in the lolling current. The sharks, apparently satiated by a previous feeding, were resting. We hovered about six feet above them for some time. They looked like tiger sharks and sharks with long tails . . . They all had names, odd names, personal names that he had given them. One in particular he called *Haku nui*, the Big Boss. He also told me of one that had been young when he was just a boy himself. As we dove down again and again, I would learn to recognize these sharks as he had . . . Sometimes, with me on his back, Kai'a would go down and come up close to the older sharks and reach out slowly with a hand to pick off barnacles that had encrusted their eyes. Such a build up of barnacles could eventually blind the old animals. They somehow trusted him and allowed him to do the cleaning. The great yellow eyes stared at us, floating inches from us as Kai'a picked away at the hard material that was often covered with *limu*. It must have hurt the sharks at least a little. They moved around slowly like a herd of cattle in a corral. Kai'a jabbed at them and pushed them away in order to stay with the shark he was working on. It was quite unreal, hanging onto this white-bearded man shoving these large, dark creatures glaring at us . . .[34]

George Augustus Robinson, a government appointee in the colony of Van Diemen's Land between 1829 and 1834, attempted to persuade the island's diminishing numbers of Aborigines to resettle on Flinders Island in Bass Strait. His controversial 'friendly mission' did not have the desired effect of Christianising and 'saving' the indigenous people from European diseases and debauchery, but his diaries have become a major source of information about the era. Regarding sharks he recorded that:

> Numbers of black women have been killed by sharks . . . When the black women see the shark in the water they dive down and remain at the bottom and cover themselves over with kelp until their formidable enemy has left . . . They say when the black man is sulky it makes the shark come to the women. It was in consequence of LULLEMOODE's being sulky (he was then quite young) and pushing the crawfish away that the shark came and killed his mother. One young woman was swimming and diving for fish when a shark came and bit at her, and caught her basket and pulled her down, when she remained at the bottom, covered herself with kelp and after some lapse of time when her aqueous region was free of danger she went to shore. Her sister had been looking at her and wondered she did not come out; thought she was killed. Asked where was her basket. Said a shark pulled her down by the basket. Some time after they found the basket and fish in it all bit to pieces. Some natives were swimming at Port Arthur when a shark bit at a woman and cut off all the fingers on one hand. She jumped on the catamaran. The sharks kept swimming round.[35]

Many European missionaries considered it their duty to destroy the so-called superstitions and ignorant beliefs of the indigenes they were 'civilising'. The sharks of the tropical

Pacific were a particular target because shark worship, in one form or another, was the *de facto* religion of many of the peoples of New Guinea, Micronesia, Melanesia and Polynesia. Early in the twenty-first century a young Canadian journalist, Charles Montgomery, travelled extensively across Melanesia, revisiting the islands where his missionary great-grandfather had proselytised in an earlier century. Montgomery found evidence that the traditions of shark worship remain, albeit under constant threat. In his quest to speak to a keeper of a shark stone—a thing of real power to the people of Vanuatu and the Solomon Islands—he was told that

> . . . the men who keep the *kastom* stones are scared. Today, I went to see the man *blong* shark stone, to see if he could bring you a shark in the bay. He refused. I told him that you would write a *bigfala* story about him . . . But he was afraid *tumas*. He said if he played with the stone, the *tasiu* [Christian priest] would kill him . . . The church wiped all the sacred fish from Langa Langa [Lagoon] . . . It's *tabu* for people to try to talk to sharks. Dangerous for the soul. We are Christians now.[36]

The demise of traditional societies through aggressive European intervention is well symbolised in the case of The Shark King versus the French Empire. The Republic of Benin, a small country on the Slave Coast of West Africa, had for centuries been the powerful economic, military and cultural kingdom of Dahomey. In the late nineteenth century colonial expansionism erupted in an orgy of greed, and the African continent was arbitrarily carved up into colonies—British, French, Italian, Belgian, Portuguese and Spanish—and shared around at a Berlin conference.

Crown Prince Kondo of Dahomey became king in 1889,

on the death of his father, Glele. The Fon people of Dahomey believed that their kings were akin to gods. On his accession, a king assumed a symbolic power-name to reflect his divine mission. Kondo saw it as his duty to keep Europeans out of Dahomey and his symbolic name—Gbehanzin Hossu Bowelle—can be translated as 'the Shark King', 'the Angry Shark' and 'the shark who made the ocean waters tremble'.[37] Voodoo divination was of central importance to the Fon and their high priests, who claimed a direct link to the spirit world, and they had predicted a difficult reign for Gbehanzin: perhaps the awesome power and attributes of the shark would assist him in his struggle against the French. Initially, Gbehanzin inflicted heavy defeats on the French (tales are still told of his heroic female military brigades), but within five years his reign was over and he died in exile in Algiers. His defiance of colonial might earned Gbehanzin the Shark King his place in the pantheon of great African leaders.

Finally, in respect of—and for—such traditions, comes this tribute and lament, written by Glenys Köhnke and published in 1974 (the year of the publication of Peter Benchley's novel *Jaws*):

> Today, to the best of my knowledge only one village remains where sharks are called and snared. The village is on the remote west coast of New Ireland. There exists in this village only one old man who knows all the magic and secret rites pertaining to shark fishing. He is very old. He has witnessed the colonising by the Germans, the war with Japan, and the administration of the Australians. As is traditional, he has passed on much of his knowledge to his nephew. This nephew has no one to whom to pass the knowledge. All his sister's sons have left the village

and gone to live in other parts of Papua New Guinea. It is more than probable that the traditions of this last remaining shark calling village will also be lost except for the record of them in this book and others in which shark fishing is mentioned . . . The shark callers have a superb knowledge of the sea and its creatures, especially sharks. It is sad that the shark callers and their knowledge and traditions are disappearing on the eve of the Western world's awakening awareness of sharks and their habits.[38]

5

'THIS STRAUNGE & MERUEYLOUS FYSHE'

Sharks and Europeans

Third Witch:
Scale of dragon, tooth of wolf,
Witches' mummy, maw and gulf
Of the ravin'd salt-sea shark,
Root of hemlock digg'd ' i' the dark,
Liver of blaspheming Jew,
Gall of goat, and slips of yew
Silver'd in the moon's eclipse,
Nose of Turk and Tartar's lips,
Finger of birth-strangled babe
Ditch-deliver'd by a drab,
Make the gruel thick and slab:
Add thereto a tiger's chaudron,
For the ingredients of our cauldron.
All:
Double, double toil and trouble;
Fire burn and cauldron bubble.[1]

William Shakespeare was hardly the first European to equate sharks with grim tidings. He wrote *Macbeth* in about 1605. One of the earliest known European depictions of a shark is on a piece of pottery found on the island of Ischia near Naples; dated to c. 725 BC, the shark is shown devouring a shipwrecked fisherman. Ancient Greek mythology tells us that Zeus, father of the gods, fell in love with Lamia, daughter of Poseidon who was god of the sea. This so angered Zeus' wife Hera that she stole Lamia's children, driving Lamia mad and turning her into an ugly sharklike creature who became addicted to eating children. (The Greek word *laimos* is translated as 'gullet'.) So distorted was Lamia's shark-face that Zeus, out of pity, gave her the power to take out her own eyes so as not to see her reflection. The word 'lamia' is still used in some Mediterranean languages to refer to the great white shark, which has in the past been classified as *Carcharias lamia*. It is also the root word for the lamnid family of sharks.

Lamia also spawned Skylla, a beautiful young woman transformed by malign fate into a sea monster with six shark-heads, who lived high up in a sea cave and regularly devoured sailors. In the words of Homer's narrator Circe:

> Therein dwells Skylla, yelping terribly. Her voice is indeed but as the voice of a new-born whelp, but she herself is an evil monster, nor would anyone be glad at sight of her, no, not though it were a god that met her. Verily she has twelve feet, all misshapen, and six necks, exceeding long, and on each one an awful head, and therein three rows of teeth, thick and close, and full of black death.[2]

Such brutally negative press wasn't universal, however. In the fourth century BC the Greek Sicilian Archestratus, considered by many to be the father of gastronomy, scorned those who were afraid to sample the culinary delights of elasmobranchs. 'There are not many men who know how heavenly is this dish or who accept to taste it. Those men are stupid like gulls and are paralysed because they say that dogfishes eat men . . .'[3] Archestratus also had a recipe for the electric ray: 'Stew it in oil, wine, fragrant herbs, and a little grated cheese'.[4]

At about this time Aristotle was compiling his monumental *The History of Animals*. He closely observed elasmobranch mating behaviour and correctly noted that the young are nurtured in eggs or live in the womb—not that he got everything right. Sharks don't breed every month, nor do any species have their young swim into the mother's mouth for safety (apart from the likelihood of their being eaten, there is no parental care of live young). But Aristotle wasn't as exaggerated in his descriptions as Pliny the Elder, whose *Natural History*, written some 400 years later, referred to whales covering three acres and sea turtle shells so large they were used as roofs and boats. Pliny's record of the ray family is closer to the mark:

> The sting-ray acts as a freebooter, from its hiding place transfixing fish passing by with its sting, which is its weapon; there are proofs of this cunning; because these fish, though the slowest there are, are found with mullet, the swiftest of all fish, in their belly . . . There is nothing in the world more execrable than the sting projecting above the tail of the sting-ray which our people call the parsnip-fish; it is five inches long, and kills trees when driven into the root, and penetrates armour like a missile.[5]

An English medieval writer, Lawrens Andrewe, wrote beguilingly of the angel shark:

> Squatinus is a fisshe in the se, of fiue cubites longe: his tayle is a fote brode, & he hideth him in the slimy muddle of the se, & marreth al other fisshes that come nigh him: it hath so sharpe a skinne that in som places they shaue wode with it, and bone also/on his skinne is blacke short here [hair]. The nature hathe made him so harde that he can nat be persed with nouther yron nor stole.[6]

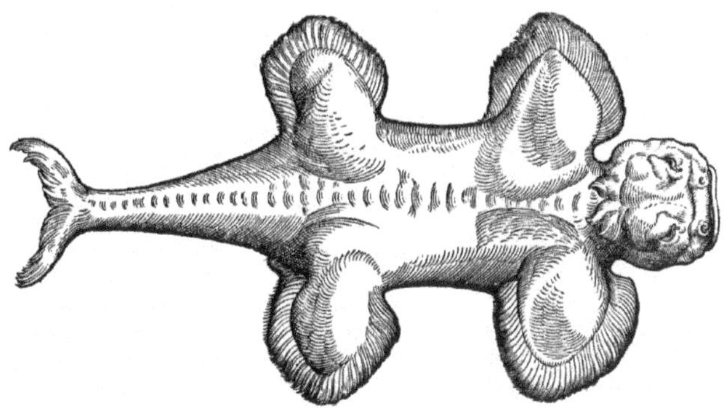

This illustration of an angel shark first appeared in Konrad Gesner's Historia Animalium Liber IV, *published in 1558.* (From *Curious Woodcuts of Fanciful and Real Beasts: A selection of 190 sixteenth-century woodcuts from Gesner's and Topsell's natural histories,* Dover Publications, 1971)

Andrewe also described the ability of numerous shark and ray species to evert their stomachs in order to eject unwanted matter: 'Scolopendra is a fisshe/whan he hathe swalowed in an angle [hook], then he spueth out al his guttes till he be quyt of the hoke/and than he gadereth in all his guttes agayne'.[7] (The scolopendra was a sea-monster in ancient Greek myth; the genus of the same name is a venomous sea centipede.)

The blurring of legend and fact is curiously persistent in the European shark story. In 1555, Olaus Magnus published his *Historia de gentibus septentrionalibus*. It became the sixteenth-century equivalent of a bestseller, was translated into numerous languages and reprinted regularly over the next hundred years. It is still a reference work of sorts. Magnus combined Swedish history and folklore, but he was blown off course in his deadly serious descriptions of Nordic sea life, describing what he claimed to be

> . . . the cruelty of some Fish, and the kindness of others. There is a fish of the kind of Sea-Dogfish, called Boloma, in Italian, and in Norway, Haafisck, that will set upon a man swimming in the Salt-Waters, so greedily, in Troops, unawares, that he will sink a man to the bottome, not only by his biting, but also by his weight; and he will eat his more tender parts, as his nostrils, fingers, &c., until such time as the Ray come to revenge these injuries; which runs thorow the Waters armed with her natural fins, and with some violence drives away these fish that set upon the drown'd man, and doth what he can to urge him to swim out. And he also keeps the man, until such time as his spirit being quite gone; and after some days, as the Sea naturally purgeth itself, he is cast up. This miserable spectacle is seen on the Coasts of Norway when men go to wash themselves, namely, strangers and Marriners that are ignorant of the dangers, leap out of their ships into the sea. For these Dogfish, or Boloma, lie hid under the ships riding at Anchor as Water Rams, that they may catch men, their malicious natures stirring them to it.[8]

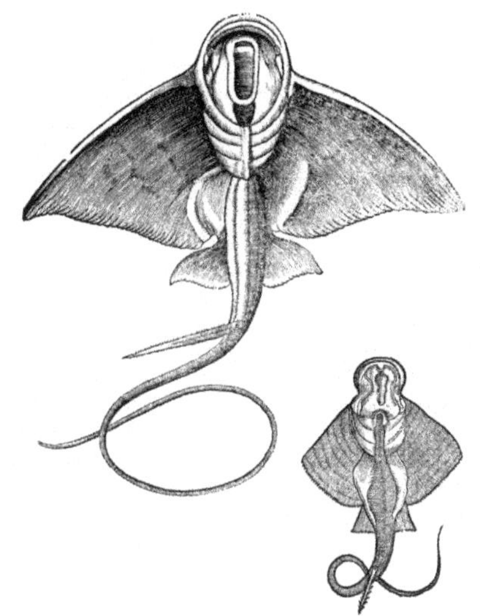

German narturalist Konrad Gesner was a contemporary of Olaus Magnus. This woodcut of an eagle ray, with its characteristic angular disc and long whiplike tail with stinging spine, appeared in Gesner's Historia Animalium Liber IV, *published in 1558.* (From *Curious woodcuts of fanciful and real beasts: a selection of 190 sixteenth-century woodcuts from Gesner's and Topsell's natural histories*, Dover Publications, 1971)

Others followed Archestratus in his pragmatic appreciation of the gastronomic possibilities of sharks and rays. In his 1604 *Ouverture de Cuisine*, Lancelot de Casteau, who described himself as a merchant, and Master Cook to three princes of Liège, includes this recipe for a dogfish pie:

> Take a piece of dogfish, four fingers in size, & put it to boil for an hour, & remove the skin close, then put it full of cloves of gillyflowers [a fragrant flower] & put it into the pie [with] some salt and mint, & chopped marjoram: when the pie is well cooked take some toasted bread, & make a pepper thereon, & put therein nutmeg, cinnamon, cloves of gillyflowers, that the sauce will be very thick, & cast on the pie when it is cooked.[9]

It's no coincidence that the words 'geography' and 'shark' were first recorded in English at about the same time, in 1542 and 1569 respectively. They were created out of necessity, to explain new phenomena that were part of the great European rebirth, the Renaissance, which encouraged the broadening of both mind and horizon. The development of printing made possible the wide dissemination of long-forgotten texts of classical antiquity, hitherto guarded jealously in monasteries or banned by successions of Popes. The Renaissance-driven scientific revolution is popularly dated to the 1543 publication of Nicolaus Copernicus' *On the Revolutions of the Heavenly Spheres.* Quite apart from the economic benefits of scientific and geographical discoveries, there arose 'a new intellectual approach to the world of nature'[10] among laymen as well as scientists, including those who sailed to the exotic new worlds. 'What was offered did not have to be part of a grand scheme of meaning. It was enough that it be "interesting", unusual or novel.'[11]

Thus it was that a large sea creature, displayed in London in 1569, caused great excitement, not least through being linked with the slave trader and freebooter Captain (later Sir) John Hawkins, who by this date had made three trips to the Caribbean region, exchanging his 'black ivory', slaves, for much wealth. Hawkins was never averse to plundering another captain's ship, and as long as his exploits brought financial rewards to Queen Elizabeth, she turned a diplomatic blind eye to his less lawful pursuits.

Hawkins' mysterious creature had no name in English. Until then, the small shark species found in inshore British waters were known as either nursefish or dogfish—both words first recorded in English at the end of the fifteenth century. (Sharks are also likened to dogs in the Italian *pescecane* and the earlier Latin *canes in mare*—both of which can be translated as 'sea dogs'. In English, 'sea dog' was the name for seals, sharks and also mariners.) It may seem surprising that northern Europeans did not apparently know about the large sharks that lived in their own waters or—in considerable numbers—in the Mediterranean. One possible reason is that there had been very little deep-sea fishing: in Britain, the home of the English language, freshwater fishes and eels had been the aquatic food of choice until about the eleventh century, when they gradually became associated with the upper classes. Only then were peasants obliged to turn to inshore and then offshore marine fishing to supplement their diets.

A London printer's broadside advertising the mysterious sea creature stated, 'Ther is no proper name for it that I knowe, but that sertayne men of Captayne Haukinses doth call it a "sharke"'.[12]

It is at this point that the waters become muddy. Where did Captain Hawkins's men get the word 'shark', and how had this particular specimen arrived in London? Several European language origins have been suggested for the word: the Anglo-Saxon *sceran* translates as 'cut' or 'shear'; the German *Schurke* translates as 'villain' or 'scoundrel'; and in Austrian German, a sturgeon is *Schirk*. Plausible though each of these may seem, all are rejected in preference to an old Mayan word *xoc* which, despite its spelling, is pronounced like 'shark' and has been linked to the presence of bull sharks in the freshwater systems of the Yucatan Peninsula.

Most sources, including the 4116-page *Compact Edition of the Oxford English Dictionary*, the respected Florida Museum of Natural History, and the *Online Etymology Dictionary*, agree that the word was brought back from the Caribbean by Captain Hawkins.[13] But what about the fish? According to the *OED*, 'the word seems to have been introduced by the sailors of Captain (afterwards Sir John) Hawkins's expedition, who brought home a specimen which was exhibited in London in 1569'.[14] The other sources named here also state that Hawkins arrived in London with the specimen.

This implication, that the Hawkins expedition might actually have returned to London with a shark, is incorrect. The creature was a thresher shark *(Aliopas* spp.*)*, the incredible tail of which is as long as the entire body. It had been inadvertently netted by English fishermen in the Strait of Dover and once in London, 'its skin was stuffed and mounted at one of the Fleet Street taverns, the Red Lion. Within a week the broadside announcing the capture and display of the fish circulated widely through the city'.[15] The lengthy description accompanying the broadside's woodcut illustration is testament to the wonder it caused (and the fluid nature of sixteenth-century written English):

The true discripcion of this marueilous straunge Fishe, whiche was taken on thursday was sennight, the .xvi. day of June, this present month, in the yeare of our Lord God. M.D.L.xix.

A declaration of the taking of this straunge Fishe, with the length & bredth, &c.

Imprynted at London, in Fleetstreate, beneathe the conduit, at the signe of Saint John Euangelist, by Thomas Colwell.

This woodcut by a London artist is of the thresher shark that was displayed in the city in 1569, attracting the first known written use of the word 'shark' in the English language. (<www.mesoweb.com/pari/publications/RT07/211-Xoc.html>)

The true discripcion of this marueilous straunge Fishe, whiche was taken on thursday was sennight, the .xvi. day of June, this present month, in the yeare of our Lord God. M. DLX. ix. A declaration of the taking of this straunge Fishe, with the length & bredth. &c. DOing you to vnderstande, that on thursdaye the .xvi. daye of this present month of June, in ye yeare of our Lord God. MD. LV. ix. This straunge fishe (whych you see here pict-tured) was taken betweene Callis, and Douer, by sertayne English Fissher men, whych were a fyshynge for mackrell. And this straunge & merueylous Fyshe, folowynge after the scooles of Mackrell, came rushinge in to the fisher mens Netts, and brake and tore their nettes marueilouslie, in such sorte, that at the fyrst they weare muche amased ther at: and marueiled what it should bee, that kept suche a sturr with their Netts, for they were verie much harmed by it, with breking and spoyling their Netts. And then they seing, and perceiuyng that the Netts wold not serue by reason of the greatnes of this straung Fishe, then they with such instruements, ingins, & thinges that they had: made such shift that they tooke this straung Fishe. And vppon fridaye the morowe after brought it vpp to Billyn-ges gate in London, whyche was the .xvii. daye of June, and ther it was seene and vewid of manie which marue-iled much at the straungnes of it. For here hath neuer the lyke of it ben seene: and on saterdaye, being the .xviii. daye, sertayne fishe mongers in new Fishstreat, agreeid with them that caught it, for, and in consideracion of the harme, whych they receiued by spoylinge of ther Netts, and for their paines, to haue this straunge fishe. And hau-inge it, did open it and flaied of the skinn, and saued it hole. And adiudging the meat of it to be good, broyled a peece and tasted of hit, and it looked whit like Veale when it was broiled, and was good & sauerie, (though sumwhat straung) in the eating, and then they sold of it that same saterdaye, to suche as would buy of the same, and they them selues did bake of it, and eate it for daintie: and for the more serteintaintie and opening of the truth, the good men of the Castle and the Kinges head in new Fish streat, did but a great deale & bakte of it, and this is moste true.

THis straunge Fishe is in length .xvii. foote, and .iii. foote broad, and in compas about the bodie .vi. foote. and proporcioned as you see here by this picture, and is round snowted, short headdid as you see, hauing .iii. ranckes of teeth on eyther iawe, maruaylous sharpe and very short .ii. eyes growing neare his snout, & as big as a horses eyes, and his hart as big as an Oxes hart, & like wyse his liuer and lightes bige as an Oxes, but all the garbidge yt was in hys bellie besides, would haue gone in to a felt hat. Also .ix. finns, & .ii. of the sormost bee .iii. quarters of a yeard longe from the body: & a verie big one on the fore parte of his backe, as you see h[. . .] by this picture, blackish on the backe & a litle whitishe on the belly, a slender tayle, and had but one bone & that was a great rydge bone runninge a longe his backe, from the head vnto the tayle, and had great force in his tayle when he was in the water. Also it hath .v. gills of eache side of the head, shoing white as you see. Ther is no proper name for it that I knowe but that sertayne men of Captayne Haukinses, doth call it a Sharke. And it is to bee seene in London, at the red Lyon, in Flete streete. Fininis.

Imprynted at London. in Fleatestreate, beneathe the

> conduit, at the sign of Saint John Euangelist, by Thomas Colwell[16]

The name 'shark' doesn't seem to have caught on quickly: for about another hundred years the English would also refer to sharks as *tiburones*, the Spanish word for 'large sharks', a word which Spanish sailors had themselves brought back from their Caribbean voyages. But negative connotations were associated with the new English word soon enough, as this 1628 definition of a social outcast shows:

> A sharke. Is one whom all other meanes have failed, and hee now lives of himself. He is some needy cashir'd fellow, whom the World has flung off, yet still claspes againe, and is like one a drowning, fastens upon any thing that's next at hand. Amongst other of his Shipwrackes hee has happily lost shame, and this want supplies him. No man puts his Braine to more use than he, for his life is a daily invention, and each meale a new Strategem.[17]

European ocean exploration generated an exponential increase in knowledge. The realisation that there were new worlds, and seas surrounding them, meant a reassessment of many prevailing certainties, in particular that human beings occupied the centre of the universe. Nature in all its forms—water, soil, rocks, fossils, plants and living creatures—came under renewed investigation. One of the scientists whose discoveries resulted in 'the essential change to the modern from the medieval world'[18] was Carl Linnaeus (1707–1778), the inventor of modern taxonomy, the classification of organisms—although it would be a long way into the future before the greatly varied elasmobranchs were arranged into their present-day orders, families and species. Back then, sharks were, literally, out of sight out of mind. But they did play a curious role in another unrelated and critically important scientific breakthrough during that dynamic period. It is in the story of *glossopetrae*, a story which also neatly symbolises the great gulf between medieval and modern.

Glossopetrae is a Latin compound word usually translated as 'tongue stones'. They were hard, pointed, pebble-like objects, commonly found on or in the ground, but they were not, however, stones. Known since antiquity, their origin had never been established. Pliny the Elder thought that they fell to earth on particularly dark nights; others believed them to be shards of lightning bolts; yet another theory was that they were nature's response to a miracle worked by St Paul when he was shipwrecked on the island of Malta:

> And when Paul had gathered a bundle of sticks, and laid them on the fire, there came a viper out of the heat, and fastened on his hand. And when the barbarians saw the venomous beast hang on his hand, they said among themselves, No doubt this man is a murderer, whom, though he hath escaped the sea, yet vengeance suffereth not to live. And he shook off the beast into the fire, and felt no harm. Howbeit they looked when he should have swollen, or fallen down dead suddenly: but after they had looked a great while, and saw no harm come to him, they changed their minds, and said that he was a god.[19]

St Paul's miraculous escape spurred him to render Malta's snakes harmless—the biblical explanation for why the island has no indigenous venomous snake species. Nature's supposed response to the saint's act was the spontaneous production of untold numbers of desiccated serpents' tongues, strewn thickly over the island's low hills and fields. Whatever their origin, *glossopetrae* were always thought to have potent magical and healing powers, and were frequently worn as personal talismans. Who would be the first to discover that these Europeans, like so many 'primitive' peoples of the largely uncharted southern hemisphere, were adorning themselves with fossilised shark teeth?

In October 1666 a young Danish scientist, Nicolaus Steno (also known as Stenson), was in the employ of the Grand Duke of Florence. When a great white shark was caught off the coast, the Grand Duke called for its head to be brought to the city for examination. The chance arrival in Florence of a huge shark's head, 'and its dissection by a young scientist eager to prove himself before a prestigious Italian court mark the unlikely beginning to an intellectual revolution that, in its way, was as profound as that of Galileo and Copernicus'.[20]

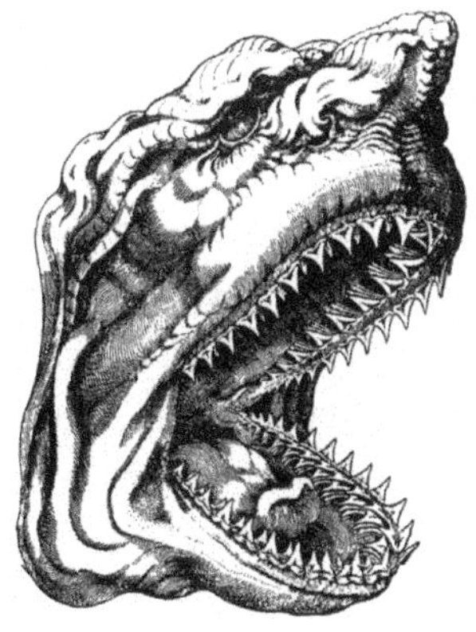

In 1667 Nicolaus Steno published Canis carchariae dissectum caput *('A shark-head dissected') advancing his theories on fossils. He used this illustration, from an unpublished catalogue of the Vatican's natural history collections, to help prove that* glossopetrae *were in fact fossilised sharks' teeth. Note the inner rows of teeth waiting to become functional.* (From *The Seashell on the Mountaintop*, Alan Cutler, Heinemann, 2003)

Steno, who happened to have had access to a collection of *glossopetrae* in Copenhagen, and had studied them in detail, was struck by the similarity between the stones and the teeth of the great white shark. While not the first to make this observation, Steno was so certain that *glossopetrae* and shark teeth were one and the same that he became determined to discover how it could be that shark teeth could be found far from the sea. This highly praised anatomist, who had already made numerous discoveries about mammalian physiology, put down his scalpel and began to travel. As he came across more and more inland marine fossil sites his quest became more complex and abstract. As well as the shark teeth, how had seashell-like objects become embedded in mountain rocks far from the sea? How could a solid come to be enclosed in another solid? Were rocks once soft? Were they originally water-carried sediments which gradually hardened around forms trapped in them?

In 1669 and 1671 Steno published two short works in which he set out the fundamentals of geology, including stratification (formation of rock or sediment layers), deep time (the Earth is far, far older than the 6000 years stated by the Bible) and superposition (the oldest rock layer is at the bottom).[21] Steno went so far as to propose that an island such as Malta may once have lain beneath an ancient sea; an audacious suggestion at the time that would be proved correct by later generations of geologists and is, of course, how Malta came to have its innumerable 'glossopetrae'. (To the dismay of his scientific colleagues, Steno subsequently converted to Catholicism. Now, three hundred years after his death, the first proponent of deep time is in the process of being beatified as a prelude to canonisation.)

As Steno was patiently tramping the hills of Tuscany, on his way to becoming the first true historian of the planet Earth, a young Englishman with no less a thirst for knowledge set out to travel the world, although he had more in common with swashbuckling John Hawkins than studious Nicolaus Steno.

William Dampier, buccaneer and navigator, would be the first man to circumnavigate the globe three times. His diaries recorded his adventures and were subsequently published as books, one of which, *A Voyage to New Holland*, published in 1703, became the most influential travel book since the publication in 1300 of Marco Polo's *Travels*.

Dampier epitomised the spirit of the age. Modestly educated, he was determined to seek out new information, which he did in abundance, recording important details about the oceans, weather, winds, animals, plants and peoples he encountered. He discovered and explored parts of Australia and he named Shark Bay in Western Australia, his ship *Roebuck* anchoring there for a week in 1699 while the crew searched the shore for provisions. His journal entries record a variety of fish, birds and mammals, including the sharks that gave the bay its name. Here is his description of a tiger shark (to him, an unknown new world shark):

> 'Twas the 7th of *Aug*. when we came into *Shark's* Bay; in which we anchor'd at 3 several Places . . . Of the Sharks we caught a great many, which our Men eat very savourily. Among them we caught one which was 11 Foot long. The Space between its 2 Eyes was 20 inches, and 18 Inches from one Corner of his mouth to the other. Its Maw was like a Leather Sack, very thick, and so tough that a sharp Knife should scarce cut it: In which we found the Head and Bones of a *Hippopotomus* [dugong]; the hairy Lips of which were still sound and not putrified, and the Jaw was also firm, out of which we pluckt a great many Teeth, 2 of them 8 Inches long, and as big as a Man's Thumb, small at one End, and a little crooked; the rest not above half so long. The Maw was

> full of Jelly, which stank extremely: However I saved for a while the Teeth and the Shark's Jaw: The Flesh of it was divided among my Men; and they took Care that no Waste should be made of it . . . And thus having ranged about, a considerable time, upon this Coast, without finding any good fresh Water, or any convenient Place to clean the Ship, as I had hop'd for: And it being moreover the heighth of the dry Season, and my Men growing Scorbutick for want of Refreshments, so that I had little incouragement to search further; I resolved to leave this Coast, and accordingly in the beginning of *September* set sail towards *Timor.*[22]

As quickly as European explorers travelled around the globe, the newly discovered species were exploited for commercial gain: the shark, for example, morphed from sea monster to Enlightenment fashion accoutrement. Commercial competition between the Dutch and British East India Companies, both vying for the spices and woods of the fabled Orient, led to an increased European interest in the East. In 1661, a Dutch newspaper, the *Hollantsche Mercurius*, reported that 900 rayskins had been imported by the Company for the Amsterdam Chamber of Commerce. Cured shark and ray skin was called shagreen (from the Persian *saghar*, the cured hide of a horse or donkey) and had been known in Europe for almost a century. The Musée de l'Armée in Paris holds a Polish Hussar's sabre with a shagreen-covered hilt which has been dated to 1559. England and France became centres for the manufacture of shark- and rayskin-covered products, which remained popular until the nineteenth century. A craze developed for covering scientific instruments—binoculars, microscopes and the like—with shagreen, as if to link the 'new' shark with the explosion of new factual discoveries. Some domestic items were also covered in shagreen, which itself was often dyed green.

The trade became so profitable that the English manufacturers kept both their sources and species of shark and ray skins closely guarded commercial secrets. In about 1760, a London shagreen casemaker, John Folgham, was advertising his business as:

> Makes & Sells all sorts of Shagreen Nurses, Fish Skin and Mahogany Knife Cases, Shaving & Writing desks, in Mahogany or Fish Skin of different Sorts, Smelling and Dram Bottles & Cases, Canister Cases &c, in Blue or Green Dog skin; Mounted in silver or plain . . .[23]

European occupation of the Australian continent also gave rise to many forms of enterprise. In December 1830, George Augustus Robinson recorded in his diary a most unusual fate befalling a group of European sealers, hardy men who worked their trade on the islands of Bass Strait, the unpredictable body of water separating Van Diemen's Land from mainland Australia. This drama occurred at Clarkes Island reef:

> They had anchored off the reef, but it came on to blow a gale of wind. The sea was running very heavy, but as the seal was going up very fast upon the rocks they kept off, expecting the wind would abate and that they should get a good knock-down. Instead of the wind abating it kept increasing, and the boat parted her cable and went on the rocks and was presently dashed to pieces. They all five got upon the reef, but they had nothing to subsist upon but to kill seal and drink the blood. Two named John Williams and John Brown, finding that the boat did not come to their assistance from Penguin Island, constructed a canoe

of seal skins by sewing them together. These rocks are about five miles from Clarkes Island and about the same distance from Penguin Island, and there is an exceeding bad tide rip, equal to the potboil off the Capisheens. They finished their canoe and next morning put to sea, John Williams, who it appears was a great reprobate, saying with a dreadful oath that was the last drop of seal's blood he would drink on that rock. The men on the reef saw them drifting away with the tide and at last entered a rip, and they never saw them more. They was probably swamped in the rip; or their canoe may have been eaten by sharks, for Parish said that some men on the Hunter [island] covered the frame of a whaleboat with the skins of seals, and was crossing to the other island when the boat was attacked by sharks, and they would soon have eaten through the seams and the whole of them must have perished, but it happened they had some carcases of wallaby kangaroo on board and they kept feeding them until they got over. It is an awful circumstance that Williams and Brown should die in their sins like this. These men had warnings sufficient to make them abandon their wicked course of living, but all the warning in the world is of no use if the grace of God does not reach their heart.[24]

For many, the 300-year Scientific Revolution culminated in the 1859 publication of Charles Darwin's *The Origin of Species*, so this commentary on sharks is included, although it constitutes a peculiarly bizarre and surprisingly unscientific observation. He wrote in his journal, while on HMS *Beagle*:

One day I was amused by watching the habits of the Diodon antennatus [porcupine pufferfish], which was caught swimming near the shore. This fish, with its flabby skin, is well known to possess the singular power of distending itself into a nearly spherical form . . . This Diodon possessed several means of defence. It could give a severe bite, and could eject water from its mouth to some distance, at the same time making a curious noise by the movement of its jaws. By the inflation of its body, the papillae, with which the skin is covered, become erect and pointed. But the most curious circumstance is, that it secretes from the skin of its belly, when handled, a most beautiful carmine-red fibrous matter, which stains ivory and paper in so permanent a manner that the tint is retained with all its brightness to the present day: I am quite ignorant of the nature and use of this secretion. I have heard from Dr. Allan of Forres, that he has frequently found a Diodon, floating alive and distended, in the stomach of the shark, and that on several occasions he has known it eat its way, not only through the coats of the stomach, but through the sides of the monster, which has thus been killed. Who would ever have imagined that a little soft fish could have destroyed the great and savage shark?[25]

6

'AN INCREDIBLY BOUNTIFUL CROP'

Shark Exploitation

In the 90 billion acres of ocean that girdle our crowded planet, an incredibly bountiful crop is often unharvested. That crop is fish, a food rich in protein and containing—unlike some forms of protein on land—all the amino-acids essential to the human diet. Yet, while an estimated two out of every three persons on earth are not getting even a minimum protein diet, one of nature's finest and most readily obtained sources of protein is virtually ignored. Some one billion tons of fish—about thirty times the current world catch—could be landed each year, and not from depleted fishing grounds such as the North Sea. But the technology of fishing remains for the most part on the level of primitive hunting, not on the level of modern farming. But we are awakening, at last, to the fact that more fish must be harvested to feed a famished world. In its Freedom-from-Hunger Campaign, the Food and Agricultural Organization [FAO] of the United Nations is seeking ways to catch and use more fish. And among them is the shark.[1]

Harold McCormick and Tom Allen's book *Shadows in the Sea*, from which the above quote was taken, was published in 1963 and has become a classic shark book. In the same year, the World Food Congress took place in Washington, and the FAO celebrated its twentieth anniversary, vowing to eliminate poverty. Sixty years later, and despite the technological advances sustaining the increasingly globalised world, the 'Make Poverty History' campaign was stark evidence of the ongoing battle to eliminate world poverty, to use those new technologies to 'feed the world'.

From the 1960s onwards, however, targeting the 'incredibly bountiful crop' in the world's oceans as a major food source, while ignoring the consequences of overfishing already apparent in the North Sea's Dogger Bank, has led to wholescale, unchecked commercial plunder of marine life forms. Among them are the sharks. Between 1960 and 1970 the world's commercial shark catch increased a massive 40 per cent, continued to rise steeply and peaked in 2000 at 869 544 tonnes.[2] Such a massive cull would affect any class of animals, but is particularly dangerous for sharks:

> Chondrichthyans are generally considered to be K-selected species, displaying conservative life history parameters such as relatively slow growth, late age at maturity, low fecundity and low natural mortality, resulting in limited reproductive output. These characteristics place them at risk of overexploitation and population depletion, with an inability to recover from reduced population levels once depleted.[3]

FAO data for the period 2007 to 2017 rank the top 20 shark-catching countries (in descending order) as Indonesia, Spain, India, Mexico, United States, Argentina, Taiwan,

Original exploitation: a traditional method of catching a shark. The shark is enticed through a noose with baitfish and the noose is then tightened behind the pectoral fins. (From *The Shark Callers,* Glenys Köhnke, Yumi Press, 1974)

Malaysia, Brazil, Nigeria, New Zealand, Portugal, France, Japan, Pakistan, Iran, Peru, Korea (Rep. of), Yemen, Ecuador.[4] By way of comparison, Indonesia's average annual catch is 110,737 metric tonnes; Spain 78, 443; India 67,391; thereafter reducing steadily to Ecuador, 7609 metric tonnes. Notable is that the UK and Canada produce just 1 per cent of global catch between them, their catches reducing significantly due to more restrictive fisheries management measures.[5]

Surprisingly just one African country appears in the table, given the severe malnourishment afflicting parts of that continent. Financial assistance to African countries has not extended to the development of modern fishing fleets, because that would not offer multinationals a return comparable to that of oil or mineral exploitation. Although China does not appear on the above list, in terms of total fisheries and aquaculture production—including teleosts, molluscs, crustaceans and aquaculture—China is by far the biggest global player, accounting for about 35 per cent of the 2020 total, followed by India and Indonesia, c. seven per cent each.[6]

Fish farming is a major industry, offsetting to some extent the plundering of wild marine stocks, but even so, unequal protective legislation in different parts of the world adds to pressures on wild marine stocks. For example, Australia's fishing fleet decreased significantly as more areas within the Australian Exclusive Economic Zone come under protection. However, there has been no decline in Australians' demand for seafood, which means an increase in imports, including from 'seriously depleted fisheries'.[7]

An explosion in fish farming in countries such as China, Indonesia, Vietnam and Thailand resulted in the clearing of the mangroves and other intertidal areas that were important shark nurseries. This affects more than the sharks; samples of seafood imported into Australia from these farms showed traces of banned antibiotics in the form of antimicrobial chemicals used in their production. One potential consequence of the continued use of antibiotics was that 'superbugs can develop and they can remain on the animal and come across to people and cause problems'.[8] The problem continues albeit in another manner:

> Intensively farmed finfish species present high mortality (50%), starting from the larval stage and continuing in sea cages. Losses are generally caused by bacterial or viral diseases; in major cases, bacterial pathologies are directly linked to the high density located in feces and sediments or to improper vaccination programs.[9]

Shark farming is not practised because of the animals' slow

reproductive and growth rates, high mortality rates in captivity, generally large size (the average shark across all species is about 1.5 metres) and the associated difficulty of keeping commercial quantities—many thousands—in close proximity. Farming a few smaller species could potentially be trialled, owing to their relatively fast growth rates. These might include the milk shark (*Rhizoprionodon acutus*), gummy shark (*Mustelus antarcticus*) and smooth dogfish (*Mustelus canis*), the latter maturing within eight years at about 150 centimetres, with litters of up to twenty pups. There is another major obstacle: 'No investor will put a huge sum of money in shark farming, only to yield limited returns a decade later.'[10]

As described in earlier chapters, sharks and rays employ a range of protective strategies, evolved over millions of years, including camouflage, dermal denticles, stinging spines, electricity, size, threat display, dark, deepwater habitats and safe, shallow pupping and nursery grounds. Since 1963, or thereabouts, these have counted for little. Modern commercial fish harvesting for human consumption is overwhelmingly thorough:

> In this type of trawl [beam trawl] the mouth or opening of the net is kept open by a beam which is mounted at each end on guides or skids which travel along the seabed. The trawls are adapted and made more effective by attaching tickler chains (for sand or mud) or heavy chain matting (for rough, rocky ground) depending on the type of ground being fished. These drag along the seabed in front of the net, disturbing the fish in the path of the trawl, causing them to rise from the seabed into the oncoming net . . .
>
> Demersal or bottom trawl is a large, usually cone-shaped net, which is towed across the seabed. The forward part of

the net—the 'wings'—is kept open laterally by otter boards or doors. Fish are herded between the boards and along the spreader wires or sweeps, into the mouth of the trawl where they swim until exhausted. They then drift back through the funnel of the net, along the extension or lengthening piece and into the cod-end, where they are retained . . .

Drift nets are not set or fixed in any way, are in fact 'mobile', and they are allowed to drift with the prevailing currents. Drift nets are used on the high seas for the capture of a wide range of fish including tuna, squid and shark . . . Despite a global moratorium on large-scale drift nets (nets exceeding 2.5 kms in length), introduced in 1992, problems still exist . . .

Purse seining [is] the general name given to the method of encircling a school of fish with a large wall of net. The net is then drawn together underneath the fish (pursed) so that they are completely surrounded. It is one of the most aggressive methods of fishing . . . Seine netting [is] a bottom fishing method and is of particular importance in the harvesting of demersal or ground fish including cod, haddock and hake and flat-fish species such as plaice and flounder. The fish are surrounded by warps (rope) laid out on the seabed with a trawl shaped net at mid-length. As the warps are hauled in, the fish are herded into the path of the net and caught. Effectiveness is increased on soft sediment by the sand or mud cloud resulting from the warps' movement across the seabed. This method of fishing is less fuel-intensive than trawling and produces a high quality catch, as the fish are not bumped along the bottom as with trawling.

Long-lining is one of the most fuel-efficient catching methods . . . It involves setting out a length of line, pos-

> sibly as much as 50–100 km long, to which short lengths of line, or snoods, carrying baited hooks are attached at intervals. The lines may be set vertically in the water column, or horizontally along the bottom. The size of fish and the species caught is determined by hook size and the type of bait used.
>
> Dynamite fishing. In some countries such as the Philippines, explosives (dynamite or blast fishing) are used on coral reefs to capture fish. Blast fishing is a particularly destructive method of fishing and is prohibited in many regions. A single explosion can destroy square metres of coral in the immediate area, whilst shock waves can kill fish in a radius of 50m or more from the blast. Reefs in some parts of South East Asia have been reduced to rubble in this way . . .
>
> Cyanide is used by fishermen in many areas of South-East Asia, the Pacific and the Indian Ocean, to stun reef fish such as grouper and Napoleon wrasse which are then exported for the live reef fish food market or aquarium trade. Although its use is prohibited the practice continues because of the demand for certain species (e.g. Napoleon wrasse) as gourmet delicacies.[11]

Official categories of shark production include frozen (whole) sharks, frozen, chilled or fresh shark fillets, sharks dried, salted or in brine, skates frozen, chilled or fresh, shark fins dried and salted or unsalted, and shark liver oil. This basic terminology can be flavoured another way: as many as 100 million sharks are manufactured into these products each year.[12]

Despite the public's psychological aversion to sharks, and despite the ammoniac taste of many species' flesh if it is not treated properly, as food products they are consumed in many ways. A random online search threw up the following international offerings: mako steak marsala; teriyaki shark steaks; baked shark cheesy surprise; grilled shark with fruit salsa; broiled spicy shark; roast angel shark; Goan stingray curry; Spanish tapa of fried tope shark in oregano marinade; smoked shark jerky; grilled shark 'to die for'; London fried skate; Cape shark in essence of fennel; Chinese baked shark; grilled shark Mexicana; shark's fin omelette.

Taste is relative. According to a 2006 article in the *Sydney Morning Herald*:

> Fear of shark attack or not, Sydney, indeed NSW, has always had a hate affair with the fish, even to the extent of declining to eat it. Maybe it is one predator's way of telling another 'we won't eat you, so you don't eat us', but the fact remains that shark always fetches the lowest prices at the Sydney Fish Market. A Bermagui commercial fisherman, Alan Broadhurst, sends his shark to Melbourne, where it fetches up to $10 a kilogram. 'But in Sydney, I'd be lucky to get $5 a kilo. Three bucks would probably be closer,' he said. The managing director of the Sydney Fish Markets, Grahame Turk, said the variety of shark species caught in NSW waters—often wobbegongs—were generally not as 'tasty' as sharks caught in the fisheries of Bass Strait and southern Australia. 'People in NSW by and large just do not go for it. But there's no doubt Victorians have developed a taste for shark,' Mr Turk said. Over the years the fishing industry has tried to rebrand shark. It was once commonly called dogfish, a name enough to put off most diners. In the early part of last century shark was renamed monkfish to cash in on the 'fish on Fridays' Roman Catholics trade,

> but evoking religion still failed to attract interest. During the Depression shark became known as blue flake and then flake, which fooled Victorians but not people in NSW. One Newcastle fisherman, who did not want to be named, said there seemed to be so many Victorians living in Sydney now that any fish and chip shop owner who renamed his business 'Flake Are Us' would make a fortune each Friday from homesick southerners.[13]

Sharks are targeted not only for their meat but also for their oil, cartilage, fins and for the aquarium trade. All shark industries also benefit from bycatch, the taking of sharks and rays in commercial fishing operations targeting other marine species. Reliable data on bycatch are limited; there is no way of knowing exactly what happens to the vast numbers of non-targeted sharks and rays hauled aboard fishing trawlers each year, but shark bycatch can be as much as 50 per cent of a commercial catch. To put this another way, half of the sharks taken out of the ocean each year are 'incidental'. Shark bycatch ranks among the most economically wasteful and ecologically damaging of all human exploitation practices. The sharks are either discarded (dead or wounded) or processed (wholly or partly). The most wasteful and barbaric bycatch practice is finning, for shark fin soup. The valuable fins are usually carved from the animal which is then thrown back into the sea, still alive, to sink and drown. Keeping the whole animal would use up hold space and also risk contaminating the commercial catch with ammonia-tainted shark meat.

Shark fins are reputed to impart certain benefits—vitality, longevity and power, all attributes of the animal itself—which is one reason why shark fin soup has always had its place in Asian, particularly Chinese, cuisine. As a delicacy traditionally enjoyed by emperors and the nobility it has considerable gourmet status and features on the menus of Chinese banquets for celebrations such as weddings and New Year festivities. In this sense, as a cultural expression as much as a commodity, shark fin soup may be likened to Beluga caviar or the best foie gras. It takes weeks of careful drying and soaking to separate out the bundle of thin collagen fibres which, supported by a cartilage platelet, grow in the fin. These fibres are called fin needles and their texture, tenderness and presentation in the soup bowl are important. The most sought-after fins are those of the blue shark, giant guitarfish (*Rhyncholsatus djiddensis*) and hammerhead sharks and they are among the costliest food products in the world, further encouraging their harvesting.

Economic and social reforms in China from the 1980s, and the emergence of a new Chinese middle class, resulted in a great increase in demand for shark fins. Documented world production of fins jumped from 2670 tonnes in 1976 to 6300 tonnes in 1997[14] and by 2006 was some 10000 tonnes.[15] This represents a fourfold increase in shark slaughter over three decades, essentially for one product, soup. CITES, the Convention on International Trade in Endangered Species, lists the great white shark, basking shark and whale shark, all three of which are exploited by the fin trade, though it is possible that their huge fins are rated more for their visual advertising potential—'trophies for display'—than their taste.[16] The oceanic whitetip, porbeagle and three hammerhead species are also at risk. Furthermore, in the words of the WWF in respect of the fin trade, 'Controls on fishing are woefully insufficient.'[17]

Sharks' fins drying in Callao, Peru, April 2007. (Oceana/LX)

A 2022 report in the *Guardian* newspaper suggests little has been done to improve fin trade management:

The International Fund for Animal Welfare (IFAW) analysed almost two decades of customs data in three major Asian trading hubs from 2003 to 2020. It found that while the main market for fin-related products is in Asia, EU countries – led by Spain (by far), Portugal, the Netherlands, France and Italy – are a significant player in supplying this legal market. China is a big supplier of shark fins but was not covered by the study . . . More than 50% of the global shark fin trade is in Hong Kong, Singapore and Taiwan province, the report says. Barbara Slee, the IFAW's report's author, said: 'Small or large, coastal or high seas, shark species are disappearing, with the piecemeal management efforts to date failing to stop their decline.'[18]

It is notoriously difficult to police the illegal taking of fins, whether from protected species or in protected waters. To take but one example, some Ecuadorean fishermen illegally cut fins from protected Galápagos Island sharks and sell them in Ecuador, where it has been legal to sell fins from sharks caught accidentally—the loophole being that it is almost impossible to prove how the shark was caught. The fin trade is attractive there because it is much more lucrative and less labour-intensive than traditional tuna fishing.[19]

In 2006, a first-time study into the shark fin trade based not on unreliable fisheries estimates but on auctions in Hong Kong, the world's main shark fin market, was published. Its findings into the 'secretive and wary' shark fin trade were depressing but not surprising. It established that shark exploitation for the finning trade was up to four times greater than an FAO estimate, with a mean total of 38 million sharks being killed per year, this figure rising to a median of 62 million (depending on species and types of fins used in the rigorously controlled statistical methodology).[20] But even that did not represent the full picture:

> In addition our trade-based biomass calculations may underestimate global shark catches. For example, due to the lack of data on domestic production and consumption of shark fins by major Asian fishing entities such as in Taiwan and Japan, unless exported for processing and then re-imported, these fins are not accounted for within our methodology . . . Furthermore, shark mortality which does not produce shark fins for the market, eg. fish mortality where the entire carcass is discarded, is also not included. These discrepancies suggest that world shark catches are considerably higher than reported, and thus shark stocks are facing much heavier fishing pressures than previously indicated.[21]

Shark fins are not the only part of the animal reputed to have curative effects on human ailments. Shark liver oil contains high levels of squalene, which is used in cosmetics and as a health product. In Australia, as elsewhere in the world, squalene is marketed as having a range of benefits, namely the temporary relief of arthritic pain, improved appearance of skin and hair, reduction in joint inflammation and swelling associated with arthritis and increase in oxygen uptake capacity. It is also claimed to reduce blood lactate levels, boost stamina and endurance, possibly reduce joint swelling and inflammation associated with gout and act as an antioxidant.[22]

Shark cartilage has been wrongly touted as a cure for cancer. The theory is that because sharks have cartilage rather than bone and sharks do not get cancer, therefore cartilage must have anti-cancer properties. (Sharks are, in fact, susceptible to a range of cancers.) Cancers require a blood supply and cartilage has no blood vessels. If cartilage has properties that prevent the formation of blood vessels, then such properties might be able to combat the development of cancerous tumours. The United States is a major producer of shark cartilage products, exporting to some 35 countries as interest in its possibilities as an alternative health product increases. Blue shark cartilage is considered to be the finest. An FAO report on non-food uses of cartilage states that:

> Many claims, not scientifically proven, attribute to shark cartilage the role of being beneficial in cases of asthma, candidiasis, eczema, allergies, acne, phlebitis, peptic ulcers, haemorrhoids, arthritis, psoriasis, diabetic retinopathy, neovascular glaucoma, rheumatism, AIDS and above all cancer.[23]

Sharks are also taken from the wild for the aquarium trade. Some small shark species such as the bamboo shark, epaulette shark (*Hemiscyllium ocellatum*) and coral catshark (*Atelomyctus marmoratus*) adapt reasonably well to captivity; others grow too large to be kept in any kind of contained environment. Some species, particularly the great white, have a history of dying when kept in captivity, although sharks are found in all major public aquaria. The Sydney Aquarium, one of the largest and most modern in the world, has numerous species including blacktip reef sharks, zebra sharks (*Stegostoma fasciatum*), lemon sharks (*Negapriou acutideus*), crested hornsharks (*Heterodontus* galeatus), Port Jackson sharks (*Heterodontus portusjacksoni*), wobbegongs, eastern fiddler rays (*Trygonorrhina fasciata*) and southern eagle rays (*Myliolsatis australis*). In 2007, for the second time, one of the aquarium's critically endangered grey nurse sharks gave birth to a single live young. Not all aquaria are accredited research facilities, however. The black market trade in sharks for the private aquarium industry is thriving.

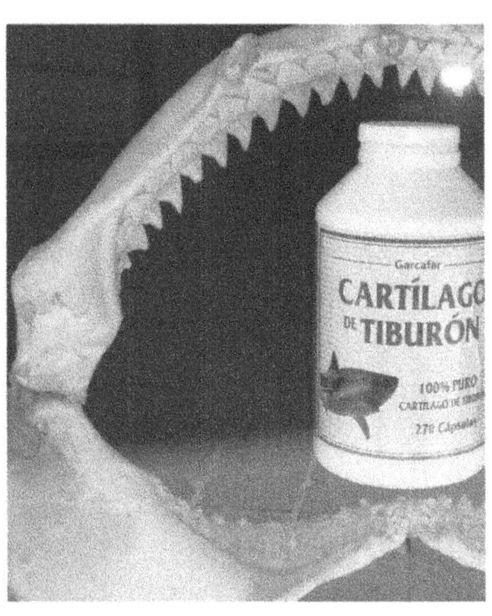

Shark cartilage in capsules and a jaw in a shop window, Madrid, Spain, 2008. (Oceana/XL.)

Sharks and rays are vulnerable to all forms of commercial fishing, by the logic of which their only intrinsic worth is their economic value:

> Fishers identified the time required to repair damaged and lost gear from shark interactions, and to remove sharks to be discarded from gear, as a substantial problem. Lost revenue from shark damage to target species can amount to several thousand U.S. dollars in a single set in some fisheries. In some of these fisheries, there is large interest in minimizing shark interactions. On the other extreme, there are pelagic longline fisheries where revenue from sharks exceeds costs from shark interactions, a large proportion of caught sharks are retained (> 99% in some fisheries), and sharks are either always an important target species, are targeted seasonally or at certain fishing grounds proximate to ports where there is demand for shark products, or are an important incidental catch species.[24]

Some commercial fishers avoid areas of high shark concentration, or move when sharks arrive in large numbers. Using fish rather than squid as bait seems to reduce shark interaction, as does deep-setting of hooks. A range of measures to minimise bycatch include new net designs, new hook types and sorting grids, which can separate out catch by size and release non-targeted fishes. Electromagnetic devices known as SharkPods may be set on longlines to repel sharks. Such innovations are, however, expensive, and the numbers of many species of sharks and rays still being caught are unsustainable.

Bottom trawl netting scoops up untold numbers of rays and benthic sharks, in the process destroying fragile seabed ecosystems. It has been described as the fishing industry equivalent of native forest clearfell: 'Trawling has a devastating effect, ripping apart delicate benthic communities and depleting fish stocks . . . Areas that are fished heavily have been reduced to bare rock, coral rubble or sand . . . Scientists are not sure how long the coral takes to recover, if at all'.[25] Midwater longline fisheries are little better, hooking vast numbers of blue sharks, as well as other predators that follow the target shoals, particularly oceanic whitetips, porbeagles, makos and hammerheads.

Recreational fishing takes its toll. Some large species of shark, such as threshers, makos and salmon sharks, are considered to be great sport to fight on rod and reel. There is a growing tendency to 'catch and release', but the 'macho' aspect of 'beating' a big shark, as immortalised by Ernest Hemingway and Zane Grey, means that the world will never be rid of photographs of proud fishermen standing next to huge, dead sharks. Then there is recreational fishing for the table. One of the longest state coastlines in the world is that of Western Australia. State government statistics indicate that

Novelist Ernest Hemingway with trophy—a large tiger shark. (John F. Kennedy Presidential Library, Boston, Massachusetts)

some 34 per cent of WA's population of two million indulges in recreational fishing, which in 2007 contributed AUD$570 million to the economy.[26] Similar statistics apply across all other Australian states.

A landmark 2006 publication, *Economically Important Sharks and Rays of Indonesia*, resulting from collaborative studies by Indonesian and Australian institutions, marked an important point at which exploitation moved from being unchecked to managed, with the next logical step being conservation. The Indonesian experience is critical, given that it is the world's fourth most populous country and, being a nation of islands, relies heavily on seafood. As noted at the beginning of this chapter, it also heads the FAO's table of top twenty shark-catching countries. The introduction to the 2006 publication states: 'Although Indonesia has the largest chondrichthyan fishery and is considered to have one of the richest chondrichthyan faunas in the world, there are almost no published biological data or size compositions of species landed'.[27] If we do not know what is being taken out of the water, we can only guess what remains—not just in Indonesian waters but globally. Ultimately, the economic benefits of harvesting sharks can never be greater than the economic benefits that will result from ensuring that they continue the role they have played for millions of years, that of keeping our seas healthy.

7

SHARK CONSERVATION

Problems, Solutions

> *Every September, one of the world's largest and densest congregations of great white sharks assembles in the waters surrounding the Farallon Islands, a 211-acre archipelago of ten islets in the Pacific, twenty-seven miles due west of the Golden Gate Bridge. No one fully understands what this gathering represents, why great whites, the ocean's most solitary hunters, choose to reside for a period of time in such close quarters . . . year after year, the same sharks return to exactly the same spot.*[1]

There is beauty in mystery, and part of the wonder of the natural world is that there is so much that we do not yet, and perhaps never will, know or understand about its inhabitants. Certainly, the great whites gather at these islands to feed off their seal colonies, but their associated socialising remains unexplained, as do their their long-distance movements, despite satellite and acoustic tag tracking. Alas, our lack of respect for many of the species with which we share this planet means that it is now imperative for us to learn as much as possible, as quickly as possible, about shark behaviour, be it that of the great white or the humble skate. If we don't, the planet's seventh mass extinction event, the one we have set in motion, will see many elasmobranch species disappear.

It is said that statistics do not lie. The following statements are taken from various conservation and government sources around the world. Their accuracy may not be uncontestable, but the twenty-first century picture they paint is bleak:

> The International Food Policy Research Institute in Washington has calculated that more than 20 million tonnes a year of fish and other marine organisms are discarded at sea. This is the equivalent of nearly 20 per cent of the world's total annual consumption of fish.[2]

> [Asia] has the world's largest fishing fleet, with 42 per cent of its registered tonnage. The Asian Development Bank says that these vessels have twice the capacity needed to extract what the oceans can sustainably produce. The result, according to the bank, is a vicious circle: as catches per vessel fall, profits plummet, and fishers overfish to maintain supplies, causing serious depletion of stocks and endangering long-term availability.[3]

> More than 125 countries around the world now trade in shark products contributing to an uncontrollable surge in the number of sharks taken from the oceans. In a little over 50 years the slaughter of sharks has risen 400 per cent to approximately 800,000 metric tons per year.[4]

> If one billion people each ate one shark fin per year, and

there are assumed to be 5 usable fins per average shark (pectorals, dorsals and caudal fin), they would consume 200 million sharks per year. Clearly the capacity for the human populace to consume sharks, much less kill them deliberately or inadvertently or to ruin their habitats, is far more enormous than the shark's ability to compensate for this by reproductive surplus, which is adjusted to natural vagaries and not an all-consuming ultrapredator.[5]

A female spiny dogfish may give birth to about 20 pups, but these need 20 to 30 years to reach sexual maturity, and pregnancy itself lasts about 22 months . . . The probability is thus very high that an animal is caught even before it has had a chance to reproduce.[6]

We [Shark Angels] headed into the field, spending years undercover documenting the heart-breaking destruction. We have walked among 7,000 bloody sharks landed in a tuna fishery in Japan, sorted through chest-high whale shark fins in a seafood mall packed with thousands of fins in China, and watched a starving fishing village in Indonesia fin the last of its baby sharks – having decimated the population in only two years. And, we found our voices, fighting for a species that needed all the help it could get.[7]

The great white shark, often referred to as the 'maneater,' with its bluish grey black and white belly remains the favourite of most shark fishermen here since it has large fins that fetch a high price. Other species of shark—hammerhead, dogfish, tiger, bluntnose sixgill and a host of others abound in Ghana's waters. 'At least we get paid between $30 and $40 for every kilogram of dried shark fins we supply, this is not bad,' Kweku Essuman disclosed with a smile. But in Singapore, Taiwan and Hong Kong, the largest centres of the shark fin trade, middlemen are paid $265 to $300 for the same kilogram of dried shark fins.[8]

Overexploitation of elasmobranchs (sharks, skates, and rays) is known to have already nearly eliminated two skate species from much of their ranges . . . Our results show that overfishing is threatening large coastal and oceanic sharks in the Northwest Atlantic . . . The trend in abundance is most striking for hammerhead sharks; we estimate a decline of 89% since 1986 . . . The trend for white sharks was an estimated 79% decline . . . Tiger shark catch rates declined by an estimated 65% since 1986 . . . The trends for oceanic sharks have also shown decline. We estimate that thresher sharks . . . have declined by 80% . . . Blue sharks declined by an estimated 60% . . . The oceanic whitetip shark declined by an estimated 70% . . .[9]

Historically, due to their fierce appearance and being mistaken for other sharks that pose a danger to humans, large numbers of Grey Nurse Sharks were killed by recreational spear and line fishers and in shark control programs, particularly in south-eastern Australia . . . The number of Grey Nurse Sharks in NSW could be as low as 292 . . . There are concerns that this population has fallen to such critically low numbers that individual animals are now failing to find mates and successfully reproduce. In addition, fishing activity, particularly recreational line fishing, is thought to be impacting severely on the existing Grey Nurse Shark population.[10]

The International Union for the Conservation of Nature (IUCN) Red List of Threatened Species is described as the world's most comprehensive source on the conservation status of plant, fungi and animal species, the purpose of which is to identify which species face an elevated risk of extinction 'in the near future'.[11] The List features many species of sharks in its various categories, along with many other species of marine life, mammals, amphibians, reptiles, birds and plants. Those in the categories Critically Endangered, Endangered, and Vulnerable are all considered to be threatened with extinction. 'Critically Endangered' means just that: as in the example above of the grey nurse shark, if a certain level of species reproduction is no longer viable then a life form may be considered to be biologically extinct. The grey nurse shark is without doubt critically endangered in eastern Australian waters; elsewhere however its numbers are less imperilled and so the species as a whole is listed only as Vulnerable. In 2023 the Australian Government Department of Climate Change, Energy, the Environment and Water estimates the east coast population at c. 2000 individuals.[12]

At the other end of the list, numerous shark species are described as 'Data Deficient', meaning that their status cannot be defined. One such example is the salmon shark, which is accepted as being common in its ranges. But this counts for little in the exploitation–conservation war. The salmon shark is

> a broad-spectrum predator that occupies a very high trophic level in subarctic waters of the North Pacific, and thus probably plays an important role in stabilizing the marine ecology where it occurs; recently, there has been a major attempt to launch a commercial fishery for this species off Alaska,

> where it does not appear to be migratory yet the basic life history parameters (population size and structure, recruitment and natural mortality rates, etc.) necessary for intelligent management of its stocks are poorly known; this seems a recipe for ecologic and economic disaster.[13]

Along with the salmon shark, those considered most at risk by the IUCN include the great white, whale shark, megamouth, basking shark, thresher, bigeye thresher, pelagic thresher, silky shark and the shortfin mako shark, Ganges shark, Borneo shark, speartooth shark, blacktip reef shark, tope shark, porbeagle, hammerhead shark, blue skate, barndoor skate, bowmouth guitarfish and at least ten ray species, including the spotted eagle ray. A major 2021 analysis of 1199 shark, ray and chimaera species using Red List criteria found 391 species (32 per cent) to be Critically Endangered, Endangered or Vulnerable. And that percentage increases if Data Deficient species are included.[14]

There are many complexities involved in grading life forms according to their future chances of survival. In 2007, an IUCN workshop was unable to accord the blue shark a higher threat rating, despite it being possibly the most exploited of all pelagic sharks (through finning), because the experts 'could not reach consensus that the species is threatened with extinction on a global scale'. Elsewhere it has been pointed out that most of the 1200-plus species of sharks, skates, rays and chimaeras are not listed. 'Are any of them threatened? Of course they are.'[15]

A recent initiative hopes to redress the acute problem. As reported in 2022:

> New hope for the conservation of sharks, rays and chimae-

> ras: Important Shark and Ray Areas (ISRAs) . . . discrete, three-dimensional portions of habitat, critical for one or more shark species, that are delineated and have the potential to be managed for conservation . . . one of ISRA's main goals is to attract the attention of policy- and decision-makers who design and develop these MPAs to the need of maintaining the favourable conservation status of sharks in those specific areas. Criteria have been designed to capture important aspects of shark biology and ecology and to encompass multiple aspects of species vulnerability, distribution, abundance, and key life cycle activities, as well as areas of high diversity and endemicity. With the finalisation of the ISRA Criteria, ISRAs are moving forward to assess the first region against these criteria and identify the first ISRAs in the world.[16]

Inshore recreational fishing takes an additional toll of sharks. In most developed countries, the introduction of bag limits and total allowable catches has gone some way towards fish stock management, but the reality is that takes began to decline as long ago as the 1970s, indicating a considerable fall in shark numbers, so the continued practice represents unsustainable exploitation. Defenders of recreational fishing counter-argue that well-kept records stretching back decades and 'tag and release' programs provide accurate data on species distribution, population and migratory habits, and that researchers have obtained much biological information from sharks taken by anglers at game-fishing tournaments. It's an advance on the notion that the best shark was a dead shark, but the survival rate of sharks caught and released is unknown, and the physical trauma associated with capture can't be ignored:

> Methods and techniques which should be used for releasing sharks are basically the same whether the fish is to be tagged or not. Always have good gloves and a wet towel handy and if possible a soft, shady surface on which to place the shark. When a shark is brought to the boat, either net it, or swing it on board and quickly lay it horizontally, preferably on a soft carpet or piece of sponge. Always avoid placing fish on hot surfaces since their skin may be quite prone to burning. Hold the shark firmly behind the head, and around the tail wrist, using gloves and/or a wet towel, and try to remove the hook if possible. If the hook cannot be easily removed, the line or trace should be cut as close to the mouth as possible. More often than not, hooks will eventually fall out, or pass through the digestive tract whereas attempting to remove a deeply lodged hook could damage internal organs or blood vessels. The internal organs of many species of shark tend to be rather loosely held in place by connective tissue. In the water, these organs are supported by the surrounding medium, but if the shark is lifted vertically, especially by the tail, connective tissue may tear internally. There is also the danger of straining or tearing tendons which hold the vertebrae in place . . .[17]

The outlook for deepwater shark species is bleak. As explained elsewhere in this book, high seas bottom trawling destroys both habitats and ecosystems. The catch constraints placed on licensed fishing fleets have led to large-scale commercial poaching. The nutrient-rich deep waters around seamounts and cold-water corals attract enormous quantities of valuable teleosts which are targeted by both legal fishing fleets and illegal poaching trawlers. There are tens of thousands of seamounts, supporting any number of unknown life forms, but many of them have been damaged or destroyed. Conservation regulations are not easily enforceable on the high seas. The former legal decimation of orange roughy stocks in the Southern Ocean off Tasmania and the illegal decimation of Patagonian toothfish stocks in the Southern Ocean near Antarctica both stood as grim testament to the rapidity and thoroughness with which demand was met at the expense of the marine environment. In 1997, orange roughy bottom trawling raised 1.6 tonnes of smashed coral to the surface for every hour of fishing time.[18] And the Patagonian toothfish fishery had 'reached the point of commercial extinction. In 1997 the total illegal catch of Patagonian toothfish was around 100,000 tonnes with a value of over AUD$760 million.'[19] In the Atlantic the deepwater Portuguese shark and the leafscale gulper shark were fished out in just 20 years.

The oceans' deeps are critically important environments. Over 60 per cent of the planet's surface water extends to depths beyond 1000 metres. Deepwater elasmobranchs are defined as those found at depths of 200 metres or more, their habitat being, 'on or over the continental and insular slopes and beyond, including the abyssal plains and oceanic seamounts'.[20] About half of the known species of sharks, batoids and chimaeras fall into this category, the least studied being those in the deepest reaches, but an estimated 20 per cent of deepwater species remain taxonomically undescribed. The reproduction rates of deepwater species are thought to be particularly slow, meaning that they are acutely vulnerable to overfishing—possibly to any form of exploitation at all.

Scientific research into the spatial dynamics of shark and ray populations has become a key factor in the chondrichthyan conservation effort. The more that is known about a species' movement patterns, migratory routes, nursery grounds and feeding habitats, the better the advice that can be given to organisations and authorities charged with combating excessive and illegal marine exploitation. Given that so many species are known to be under threat, it could be assumed that such research would be given top priority. Unfortunately, this is not the case, primarily because of the costs involved:

> 'Satellite tagging that provides information on actual migratory routes—rather than just start and end positions—and on depth behaviour is expensive . . . conventional tagging requires tagging large samples of each component of the population—newborns, juveniles, adult males, adult females . . . genetic studies require collection of large samples from different areas . . .'[21]

Many governments are not prepared to fund such work, the complexities (and, therefore, the costs) of which are compounded by the huge distributions of species such as the blue shark, oceanic whitetip and great white. As an illustration of the difficulties involved in studying the largescale spatial dynamics and population structures of the great white, pop-up satellite tagging in 2002 and 2003 revealed the surprising information that 'a white shark performed a previously unknown fast transoceanic return migration spanning the entire Indian Ocean, swimming coast-to-coast from South Africa to Australia and back'. The journey also involved dives to depths of nearly 1000 metres and tolerance of water temperatures as low as 3.4 degrees Celsius.[22]

Global problems require global solutions. The United Nations, through its agency the IUCN, has for many years been working to manage and protect the world's flora and fauna. Many thousands of scientists and experts are networked to the IUCN through its many subsidiary agencies, chief among which are the Species Survival Commission (SSC) specialist groups representing amphibians and reptiles, birds, invertebrates, mammals, plants and fishes. (To name but a few: the Crocodile Specialist Group; the Marine Turtle Specialist Group; the Flamingo Specialist Group; the Mollusc Specialist Group; the Antelope Specialist Group; the Canid Specialist Group; the Arabian Plant Specialist Group.) There are multiple fish groups: coral reef fish; grouper and wrasse; salmonids; sturgeon; freshwater fish; and sharks.

The Shark Specialist Group (SSG) came into being in 1991. Its membership is selected from global geographic regions.[23] In addition to working on conservation, research, management and education issues, its major goal is to carry out Red List assessments for all of the known species of chondrichthyans—a significant and challenging undertaking. The SSG also works closely with CITES and TRAFFIC, the agency monitoring the international trade in wildlife.

In 1999, the FAO published a significant resolution, the FAO International Plan of Action for the Conservation and Management of Sharks, commonly known as 'IPOA-Sharks'. The resolution comprised approximately 30 points for conservation and management actions by member states of the FAO. Some were straightforward, some complex and open to interpretation:

> States, within the framework of their respective competencies and consistent with international law, should strive to cooperate through regional and subregional fisheries organizations or arrangements, and other forms of cooperation, with a view to ensuring the sustainability of shark stocks, including, where appropriate, the development of subre-

> gional or regional shark plans . . . Where transboundary, straddling, highly migratory and high seas stocks of sharks are exploited by two or more States, the States concerned should strive to ensure effective conservation and management of the stocks.[24]

The SSG became the key body assessing the implementation of this resolution. In 2002, it reported that there had been 'negligible progress . . . largely because the IPOA-Sharks is wholly voluntary and appears not to be considered a priority by many shark fishing States or Regional Fisheries Management Organizations'.[25] Sadly, but not surprisingly, only a handful of countries have drawn up comprehensive strategies for management of their fisheries, and most of those are wealthy nations that can afford to put relatively strict controls on this area of their income generation.

IPOA-Sharks was also supposed to have been a trigger for each member nation to investigate formally the present status of its shark catch and related shark stock issues within its exclusive economic maritime zone, but very few of those investigations were completed. One which was: *The Australian Shark Assessment Report* (2002), a comprehensive 200-page document compiled for the Commonwealth Department of Agriculture, Fisheries and Forestry by a shark advisory group whose 21 participants were drawn from State Fisheries Management Authorities, Commonwealth bodies, the recreational fishing industry and indigenous, scientific and conservation bodies. It was an interesting survey of a nation ranked well down the list of shark catchers and producers (then taking less than two per cent of the annual global catch). Some facts and statistics from just 20 years ago:

- Australia's vast Economic Exclusion Zone (EEZ) extends for 200 nautical miles around the continent and to the same extent around Norfolk Island, Lord Howe Island, Christmas and Cocos Islands, Macquarie Island, Heard Island and the Antarctic waters off its Australian territories. Each of the country's six states has jurisdiction over the first three nautical miles from its shores.
- Statistics are compiled for six forms of shark take: directed (targeted), non-directed (incidental), bycatch (discarded incidental), byproduct (retained incidental), recreational (from shore, private boat, charter, game fishing) and beach protection programs.
- Statistics are unknowable for shark 'cryptic mortality' from such factors as net drop out (entangled sharks which fall out or are eaten by other sharks), ghost fishing (lost nets in which sharks can get entangled and drown), high voltage submarine cables linking electricity grids, seismic activity resulting from oil and mineral exploration, and fatal damage to caught sharks returned to the water.
- Australian waters are home to 322 known species of sharks, rays and chimaeras, half of which are endemic to Australian waters and most of which are demersal. Some 50 species are listed as being of concern.
- Sharks represented less than five per cent of Australia's annual fish catch in 1998–99, at a total of 8593 tonnes valued at AU$22.7 million.
- In the same period the gummy shark represented nearly one-third of the shark catch, followed by the school shark at just under ten per cent and the dogfish at just under five per cent. Species such as hammerheads and wobbegongs represented less than one per cent each.

Fully one-third of sharks taken were classified as unidentified.

- Queensland and New South Wales together have shark protection measures—drums, hooks and meshing—at 121 beaches. About two sharks are caught this way in Queensland each day, about one every two days in New South Wales. There has been a significant reduction in sharks taken this way since the 1970s, but it is not known if this reflects population decreases (or possibly increased shark awareness of nets).
- Australia has an estimated four million recreational anglers who fish for 50 million days per year.
- Finning is banned in Australian waters.
- Cartilage processing is increasing.
- Exploitation management processes include estimating the total biomass of a species and reducing that mass to no more than a certain percentage of that total.
- Overwhelmingly, the Australian shark fishing industry is for the fish and chip trade.
- The Game Fishing Association of Australia recognises eight species of shark for competition game fishing: blue, tiger, gummy, porbeagle, hammerhead, thresher, mako and whaler. The great white shark is no longer recognised because of its protected status.[26]

This report served as the working document for Australia's National Plan of Action for the Conservation and Management of Sharks (Shark-plan), released in 2004. Shark-plan is 'a national guide for managers and interested stakeholders on how to better incorporate shark conservation and management issues into the management of fisheries and the broader marine environment'.[27] It sets out ten Objectives, identifies eighteen Issues, six broad Themes and 43 Actions—seemingly a classic case of bureaucracy-speak, except that this is no laughing matter. The combination of species annihilation and large-scale habitat destruction makes the rapid implementation of such an action plan all the more important. Thus each Issue addresses an Objective, and is a 'need to' or a 'need for', that is, an action that can be implemented. The Actions are graded according to priority for implementation, from 'within 12 months' to 'within four years if not sooner'. The responsible agencies for each Action are clearly identified and funding issues are addressed. It is, in effect, a Business Plan for Sharks, with a committee set up to begin implementing its recommendations.

In 2012 Shark-plan 2 was released, building on the results of the first plan: 'Australia is a world leader in the ecologically sustainable management and use of natural resources. Shark-plan 2 provides a framework for the long-term conservation of Australia's shark populations, and for guiding the industries and communities that affect them.[28]

As more and more people become aware of the uncertain future facing the planet's elasmobranchs, an impressive range of organisations throughout the world have emerged, all dedicated to shark conservation. Collectively they play a major role in educating and informing the general public and lobbying governments, with a freedom perhaps denied to some of the government-affiliated agencies equally devoted to shark conservation. A representative sample of entities wholly or partly dedicated to some form of elasmobranch conservation shows that many are closely linked. This sampling can be categorised broadly by type (although there is a good deal of crossover between types): national/ international; regional; educational; advocacy; single species; single focus. The youthfulness of most of these organisations reflects the reality that not very long ago shark conservation did not exist, but it also confirms that we have come a long way since *Jaws*.

Some national and international shark conservation organisations

The *European Elasmobranch Association*, founded in 1996, coordinates the regional and international activities of numerous European organisations involved in shark conservation, management, education and research. In other words, it is an association of organisations, not individuals. At present, organisations from 13 countries are involved and the association's members meet annually. One example of its work was a major report published in 2007. The report focused on the European Union's regulation of the ban on finning and concluded that 'the current EU Shark Finning Regulation cannot be characterized as effective', setting out a raft of recommendations including a 'fins attached' policy, that is, sharks should not be processed until landed at the dock. The report urged its members to lobby EU officials to enact appropriate legislation.[29] The 2023 annual meeting took place in October in Brighton.

The *American Elasmobranch Society*, founded in 1983, has hundreds of American and international members (individuals). It is a non-profit organisation that seeks to advance 'the scientific study of living and fossil sharks, skates, rays, and chimaeras, and the promotion of education, conservation, and wise utilization of natural resources'.[30] The society's annual meetings give rise to management resolutions such as the following from 2006, in respect of the sandbar shark (*Carcharhinus plumlseus*):

> Whereas the most recent, peer-reviewed stock assessment for northwest Atlantic Ocean sandbar sharks is significantly less optimistic than the previous assessment; Whereas National Marine Fisheries Service scientists have determined that this sandbar population is in an overfished state and overfishing is occurring; Whereas Atlantic sandbar sharks are now estimated to require more than 60 years to rebuild from their current overfished state; and Whereas the sandbar shark has been the most important species in US Atlantic commercial and recreational shark fisheries; Therefore Be it Resolved that the American Elasmobranch Society urges the National Marine Fisheries Service to immediately begin the process to reduce fishing mortality on sandbar sharks by strengthening Atlantic fishing regulations.[31]

In the same year a subcommittee of the Society undertook an 'AES International Captive Elasmobranch Census', conducted in 64 aquariums in the United States and 48 in Europe, Australasia, Africa and elsewhere in the world. The census recorded just under 10000 captive individuals, comprising 103 species of ray, 86 species of shark and two species of chimaera, with by far the most common captives being the cownose ray, followed by the bamboo shark. At the other end of the scale, sixteen bull sharks were recorded, four tiger sharks and five whale sharks.

In 2021 the inaugural 'American Elasmobranch Society Global Wedgefish & Guitarfish Symposium' was held virtually over two days to discuss research, conservation and management of these mostly critically endangered 'rhino rays', the Order *Rhinopristiformes.* A total of 37 countries were represented at the Symposium, with presentation titles leaving no doubt as to the plight of these animals and the efforts to save them. Here are a few:

- Playing for time: Guitar and violin sharks – is this the last dance?
- Rhino rays – uncovering the drivers behind their exploitation in data-poor areas
- High fishing pressure and trade driving rhino rays towards extinction – a case study from the Bay of Bengal, Bangladesh
- A multi-technique approach to understanding wedgefish ecology in the Bazaruto Seascape, Mozambique
- Creating connection: Engaging zoo and aquarium visitors in the conservation of wedgefish and guitarfish.[32]

The *Commonwealth Scientific and Industrial Research Organisation* (CSIRO), Australia's national science agency, was founded in 1926 and is active across many disciplines. Its Tasmanian Hobart-based marine scientists look after one of the world's largest national marine jurisdictions. The Hobart division cooperates in many regional and international projects, including elasmobranch research. Many hundreds of great white sharks have been satellite- and acoustic-tagged, providing statistical insights into their movement patterns, behaviour, juvenile nursery areas and adult population estimates for the eastern and southern-western Australasian populations. And these results are further refined through DNA 'fingerprinting'. One white shark created history:

> The longest continuous satellite track of a white shark to date is for a 3.6-metre male nicknamed 'Bruce'. Bruce was tagged at North Neptune Island [South Australia] in March 2004 by [CSIRO] scientists with help from staff of the Mel-

> bourne Aquarium. Bruce remained around the Neptunes for several days before heading rapidly east through Bass Strait, then north along the NSW and southern Queensland coasts. He spent most of the winter in offshore waters east of Rockhampton. In late October, Bruce returned south and last transmitted from eastern Bass Strait in early November. Bruce covered a distance in excess of 6000 km during this eight-month period.[33]

An acoustic tagging program tracked gulper sharks (*Centrophorus spp.*) in deep waters off the South Australian coast, where fishing had severely depleted their numbers. In both cases, identifying a pattern of the sharks' movements enabled scientists to advise on commercial fishing activities likely to impact further on the species. Population monitoring is also undertaken with species such as grey nurse sharks and river sharks. Tagging typically operates as follows:

> Sharks are fitted with acoustic tags that emit a sequence of low frequency 'clicks' which give each tag an audible ID number. These unique signals can be detected and recorded when the shark swims within range of the shark monitoring receivers. To capture a shark a single baited hook is suspended from a large, anchored buoy that is monitored continuously. When safe to do so, experienced research staff secure the shark with suspension ropes and carefully roll the shark onto its back, putting it into a sleep-like state known as 'tonic immobility'. A small incision is then made in the shark's abdomen, the tag inserted into the body cavity and the incision closed with a few stitches. At the completion of the minor surgery the shark is rolled over. Even though the process only takes a few minutes, great care is taken to keep

> the shark's head and gills in the water so that it can continue to breathe during the procedure. A plastic identification tag is attached to the dorsal fin as a visual record that the shark has been internally tagged. Details about the shark are recorded such as species, sex, length and a genetic sample is taken. Once the tagging process is complete, lines are carefully removed and the shark is released.[34]

CSIRO in Hobart also maintains and develops the Australian National Fish Collection, with some 145000 preserved specimens including many Indo-Pacific sharks and ray species. It is not a public access collection, its primary role being as a biodiversity reference and research tool.

The *Oceania Chondrichthyan Society* is one of the newer entities, formed in 2005. The society held its inaugural AGM in August 2006, in Hobart, Tasmania, delivering a charter to include elasmobranch research, education, conservation and sustainable utilisation. It was established as a joint venture between Australia, New Zealand, Papua New Guinea and the Pacific islands of American Samoa, Cook Islands, Federated States of Micronesia, Fiji, French Polynesia, Guam, Kiribati, Marshall Islands, Nauru, New Caledonia, Niue, Northern Mariana Islands, Palau, Pitcairn Islands, Samoa, Solomon Islands, Tokelau, Tonga, Tuvalu, Vanuatu and Wallis and Futuna islands. Many new species of Australian sharks and rays were described at the September 2008 conference. Among its aims is a desire to promote 'a friendly, relaxed, egalitarian and welcoming atmosphere for members and those interested in the study of Chondrichthyes'.[35]

It is such an approach that drives collective deliberation and action, but often those who should be working together come at an issue from opposite directions, as demonstrated by a May 2007 newspaper article posted on the society's News and Events page. Headed 'Warnings over shark numbers don't add up', the article was based on an underwater photograph taken at Seal Rocks on the mid-north coast of New South Wales, of a congregation of at least 60 grey nurse sharks and rebuts the claim that the Australian east coast population of grey nurse sharks may be close to extinction:

> To have 20 per cent of the population in one place is not mathematical or logical, especially when we claim we don't know where their habitats are,' a NSW Fisheries source told *The Daily Telegraph*. 'There are much more than 500 . . . the Government has to admit it has made a mistake.' *The Daily Telegraph* understands there is internal disquiet within NSW Fisheries at the policies of the [State] Government over grey nurses. The Nature Conservation Council (NCC) is pushing for more exclusion zones in NSW fishing areas on the grounds the species is nearly extinct . . . Anglers and divers say the breed is thriving in reefs not checked by scientists . . . Underwater Spearfishermen's Association secretary Mel Brown, who has kept records of shark numbers for 10 years, believes they have grown to 6000 . . . Mr Brown said: 'The Government has not admitted there are more sharks despite anecdotal evidence and photos showing otherwise. People claim we are bringing this to light because we want to fish them. That is not the case, we are for conservation. We just don't want to be locked out of our favourite fishing spots.' Primary Industries Minister Ian Macdonald said claims the grey nurse shark was no longer an endangered species were based on pure speculation and not science.[36]

The *National Shark Research Consortium* was an important

Congress-funded start-up a scientific collaboration (2001) between four leading American institutions in the field: the Florida Museum of National History, the Mote Marine Laboratory (Florida), the Pacific Shark Research Center (California) and the Florida Program for Shark Research. The consortium worked closely with government in 'assessing the status of shark stocks, managing U.S. shark fisheries, and helping the U.S. take the leading role in worldwide conservation and management of shark populations . . . the four organizations of the NSRC are uniquely qualified to conduct leading studies of sharks and their relatives on national, international and global scales'.[37] The consortium ran concurrent projects, such as this 2007 research project into industrial contamination of freshwater sharks:

> To determine the health risks that human pharmaceuticals pose to juvenile bull sharks and other wildlife inhabiting wastewater-impacted rivers, Mote Marine Laboratory's Center for Shark Research has just initiated a new study on pharmaceutical exposure in bull sharks from Florida's Caloosahatchee River. In this study, Mote researchers will be examining the exposure of bull sharks to synthetic estrogens used in human birth control pills by tagging sharks with standard fish tags bearing passive sampling devices similar to personal exposure badges, which are commonly used to measure chemical and radiation exposure in humans. These devices are designed to accumulate environmental pollutants and, if the shark is recaptured, they can be used to examine the chemicals that an individual fish was exposed to. In addition, Mote scientists will also be measuring the uptake of synthetic estrogens and a number of other human drug-related compounds in juvenile bull sharks by measur-

PLATE 1: Prickly dogfish (*Oxynotus bruniensis*) © *Rodie Kuiter, OceanwideImages.com*

PLATE 2: Portuguese shark (*Centroscymnus coelolepis*) © CSIRO

PLATE 3: Frilled shark (*Chlamydoselachus anguineus*) © CSIRO

PLATE 4: Grey reef shark (*Carcharhinus amblyrhynchos*) © www.natoliunderwater.com

PLATE 5: Bull shark (*Carcharhinus leucas*), feeding

Plate 6: Scalloped hammerhead shark (*Sphyrna lewini*)

PLATE 7: Collared carpet shark (*Parascyllium collare*) © Rodie Kuiter, OceanwideImages.com

PLATE 8: Spotted wobbegong (*Orectolobus maculatus*) L. Conboy, © CSIRO

PLATE 9: Zebra, or leopard, shark (*Stegostoma fasciatum*) © www.natoliunderwater.com

PLATE 10: Goblin shark (*Mitsukurina owstoni*)

Museum of New Zealand Te Papa Tongarewa, negative number P.031009

PLATE 11: Shortfin mako shark (*Isurus oxyrinchus*) *© Andy Murch, OceanwideImages.com*

PLATE 12: Grey nurse shark (*Carcharias taurus*) *© www.natoliunderwater.com*

Plate 13: Great white shark (*Carcharodon carcharias*)

© Chris & Monique Fallows, OceanwideImages.com

Plate 14: Basking shark (*Cetorhinus maximus*)

© Andy Murch, OceanwideImages.com

Plate 15: White-spotted guitarfish (*Rhynchobatus laevis*) *G. Yearsley, © CSIRO*

PLATE 16: Freshwater sawfish
(*Pristis microdon*)
© CSIRO

PLATE 17: Maugean skate (Zearaja maugeana)
(*ventral view, note claspers*)
T. Carter, © CSIRO

PLATE 18: Elephant fish
(*Callorhinchus milii*)
T. Carter, © CSIRO

PLATE 19: Ancestral sawfish, depicted by Nekingaba Maminyamanja

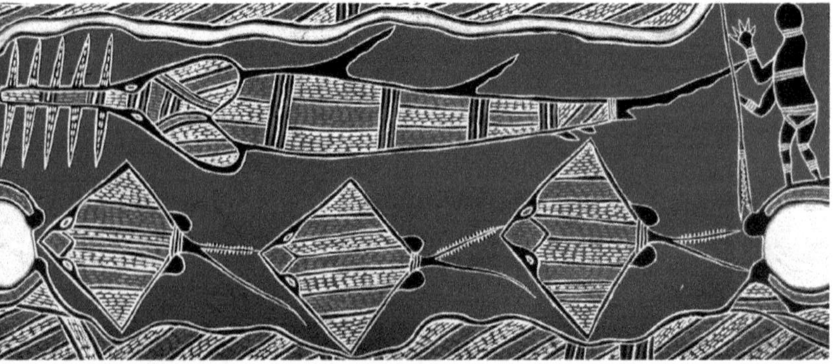

Reproduced courtesy of the artist

PLATE 20: Antoine Berjon, *Still Life with Flowers, Shells, a Shark's Head, and Petrifications* (1819)

Philadelphia Museum of Art

Plate 21: John Singleton Copley, *Watson and the Shark* (1778)
Ferdinand Lammot Belin Fund, image courtesy of the Board of Trustees, National Gallery of Art, Washington DC

Plate 22: Winslow Homer, *The Gulf Stream* (1899)
The Metropolitan Museum of Art, Catharine Lorillard Wolfe Collection, Wolfe Fund, 1906 (06.1234) © The Metropolitan Museum of Art

> ing the presence and concentrations of these chemicals in shark blood, which can be sampled using non-lethal approaches.[38]

The *Florida Program for Shark Research* is run out of the Florida Museum of Natural History. It states, with good reason, that its website 'is one of the largest and most frequently accessed elasmobranch site on the Web.'[39] As noted in Chapter 1 the Program runs the International Shark Attack File; also the International Sawfish Encounter Database (documenting all known encounters worldwide); and the Chondrichthyan Tree of Life Project (comprehensive documentation through anatomical, DNA and fossil data).

The *KwaZulu-Natal Sharks Board Maritime Centre of Excellence* claims the honour of being 'the only organisation of its kind in the world'.[40] It came into being in 1962 as the Natal Sharks Board with the (then) sole aim of protecting bathers along the Natal coastline. In choosing nets, the South Africans were following the lead of the Australians, who had started protecting their beaches in the late 1930s.

> Most of the shark nets deployed by the KZNSB are 214 m long and 6 m deep and are secured at each end by two 35 kg anchors; all have a stretched mesh of 51 cm. The nets are laid in two parallel rows approximately 400 m offshore and in water depths of 10–14 m. A drumline consists of a large, anchored float (which was originally a drum) from which a single baited hook is suspended. Most beaches are protected either by two nets or by one net and four drumlines, but the quantity of gear varies from beach to beach . . . Shark nets do not form a complete barrier and sharks can swim over, under or around the ends of the nets. Neither, of

course, do drumlines form a physical barrier. Both types of equipment function by reducing shark numbers in the vicinity of protected beaches, thereby lowering the probability of encounters between sharks and people at those beaches. The nets may have a limited barrier effect as well, but the fact that about one-third of the catch is caught on the shoreward side of the nets is evidence that such an effect is only partial . . . By 2019, the KZNSB had deployed 165 drums along the coast, reducing the length of nets to 15km (representing an almost 70% reduction) . . . The idea of introducing drumlines is to reduce the bycatch of harmless non-shark species such as whales, dolphins and turtles, which are accidently caught in the nets. The capture of non-target species has been reduced by 47,5% with the installation of these drumlines.[41]

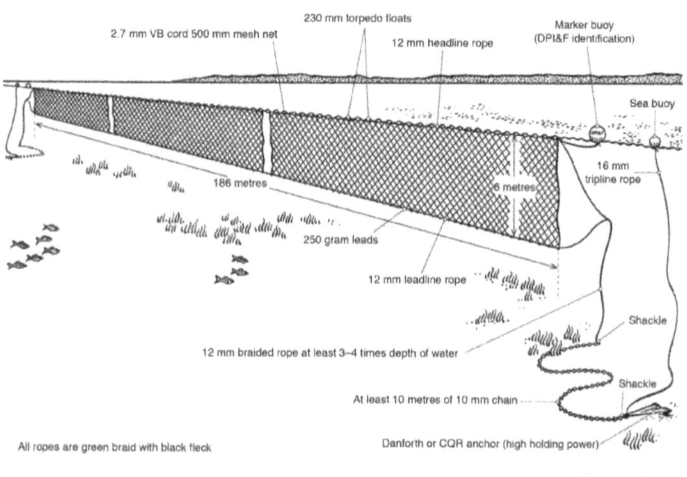

A typical configuration of shark nets. Many sharks, large teleosts and marine mammals and reptiles become tangled in nets and drown. (Queensland Department of Primary Industries & Fisheries)

Despite being a service organisation that protects beach users in KwaZulu-Natal against shark attack, the KZNSB has a strong conservation ethic and is closely networked to the global movement to protect sharks.

The US-based *Shark Conservation Fund* is a well-funded philanthropic body which

> excels at identifying and funding the most strategic projects that are focused on the global defence of sharks and rays. SCF's comprehensive approach, extraordinary human capital, and extensive international reach positions it well to influence all major shark fishing, processing, and consuming countries . . . Overfishing is killing the ocean. Outside of climate change, overfishing is the main contributor to the rapid decline of ocean health which is causing human catastrophe through loss of jobs, food security, profits, habitat and ecosystem balance. Of the earth's nearly 8 billion people, over 1 billion depend on seafood as their primary food source and the industry provides over 780 million jobs worldwide. Without immediate action, the precipitous decline of this resource will result in accelerated environmental degradation, international conflict, and hunger. Confronting the issue through shark and ray conservation is a catalytic solution to this global problem. [42]

Educational conservation organisations range in their scope. As an example, the *Bimini Biological Field Station* off the coast of Florida has a mission to

> advance our knowledge of the biology of marine animals, especially the heavily impacted elasmobranch fish fauna

> (sharks and rays); to educate future scientists at undergraduate and graduate levels; and to disseminate our research results to advance the field of marine science and conservation biology, as well as raise public perception and awareness of sharks and other marine species.[43]

The Shark Research Institute (SRI), founded in 1991 at Princeton, New Jersey, is dedicated to shark research, education, conservation and legislation. It states: 'Join us on our mission. The sharks need all of us.'[44] Today as a significant multi-disciplinary non-profit organisation it has multiple offices in the US and internationally in 11 countries including the UK, Australia, India, Mexico, Seychelles and Honduras. Its message is hard-hitting:

> The challenges facing this critical apex predator Are many, and seemingly overwhelming. More than 100 million are slaughtered every year for their fins. Climate change is dramatically disrupting the oceanic ecosystem and food chain. And mankind's attitudes are indifferent at best; savage at worst. Few species are more misunderstood, and more vilified by popular culture. We're working to change that. Through funding and sponsoring credible research, we provide fact-based, peer-reviewed information to inform. Beyond hyperbole. Beyond TV ratings. Beyond question.'[45]

Wild Oceans, formerly the *National Coalition for Marine Conservation* was founded in 1973, by a group of Virginian fishermen. In its 50 years

> we've dedicated ourselves to uniting fisherman and other ocean advocates to form an unprecedented network of passionate ocean stakeholders. Our mission is to keep the

> oceans wild to achieve a vibrant future for fishing by building coalitions and engaging in marine fisheries management using science, law and ecosystem-based solutions. We bring conservation-minded fishermen and other ocean advocates together to promote a broad, ecosystems approach to fisheries management that reflects our expanding circle of concern for all marine life and the future of fishing. Our programs emphasize conserving the ocean's top predators – the big billfish, swordfish, tunas and sharks that are the lions, tigers and wolves of the sea – while preserving healthy ocean food webs and critical habitats essential to the survival of all fish, marine mammals and seabirds. So much of what we love about the sea, about fish, about fishing, is in the wildness. But that wild world, and the future of fishing, now hangs in the balance.[46]

In 2022 Wild Oceans with other conservation organisations succeeded in implementing a two-year ban on the retention of all North Atlantic shortfin mako sharks caught by commercial or sports fishers; a ban already applying to numerous other vulnerable species. The ban applies to member states of the International Commission for the Conservation of Atlantic Tunas (ICCAT) which includes 51 fishing nations and the European Union. The intention of the ban is to remove all incentives to catch makos – the meat is valuable and for sports fishers it is 'legendary fighter' – because

> makos' conservative life history, including late maturity and low reproductive output, accelerated the overfishing problem. Shortfin mako sharks grow slowly, and females do not reach maturity until 19 – 22 years of age. Because of this,

> it is anticipated the spawning stock biomass will continue to decline for many years after fishing pressure has been reduced and until the recruits reach maturity.[47]

Advocacy for sustainable fishing is both understandable and praiseworthy, as is advocacy for the dive industry. The *Project Aware Foundation* based in New South Wales, Australia, describes itself as the dive industry's leading non-profit environmental organisation. It came into being in 1989, its name derived from an acronymn— Aquatic World Awareness Responsibility and Education. Its achievements are many. The Foundation

> Helped secure international protection policies for over two dozen vulnerable shark and ray species . . . Contributed ground-breaking science on marine debris, removing 2 million pieces of trash from the ocean . . . Created the largest and quickest growing underwater citizen science movement on the planet . . . Educated 1 million divers on the environmental threats facing the ocean and how to take action.[48]

The *Australian Marine Conservation Society*, headquartered in Queensland has a similar focus: 'We are Australia's leading national charity dedicated solely to protecting our precious ocean wildlife.'[49] Formed in 1965, the Society

> works on the big issues that risk our ocean wildlife. Together, we have protected critical ocean ecosystems with marine reserves around the nation, including Ningaloo and the Great Barrier Reef. We have led the movement to ban whaling, stopped supertrawlers, and protected threatened

> and endangered species like the Australian Sea Lion. Together, our community of ocean lovers save our oceans every day.[50]

And in respect of sharks:

> Thanks to us, live shark finning at sea is now illegal in all Australian states and the Northern Territory. The NT was the last Australian jurisdiction which allowed this cruel and wasteful fishing method, and our community campaign resulted in its ban in 2004. We continue to work towards a ban on the export and import of shark fins in Australia, to stop our involvement in this terrible trade.[51]

The Plymouth, UK-based *Shark Trust*, founded in 1997, is a highly active UK-based marine charity. It is the UK member of the European Elasmobranch Association, and also associated with other conservation groups in forming the Shark Alliance, a former coalition of some 40 non-government organisations dedicated to the conservation of sharks. That Alliance's successful core mission was to tighten up European fishing policy because of its damaging global impact on shark stocks.

One of the Trust's more unusual shark conservation projects is the successful and popular 'Great Eggcase Hunt'. Many elasmobranch species reproduce by laying eggs – as noted in Chapter 3, after fertilisation a tough leathery capsule-like sac grows around the eggs before the female releases them through the cloaca. This egg sac is often known as a mermaid's purse and can be found in a variety of shapes, some of which have tendrils to help anchor the sac to the seabed or rock. After several months a fully formed shark or skate emerges. These empty capsules frequently wash up on shores.

The Great Eggcase Hunt was initiated by the Trust in 2003, as a public initiative with a sound monitoring and conservation motive: empty cases help indicate species presence and diversity. The 2022 Great Eggcase Annual Report statistics are impressive: 372, 597 eggcases recorded in nearly 20 years; 49 species recorded; 8,170 people have submitted eggcases from 30 countries. As a young person-friendly activity it is laudable and increasingly popular, with invariably interesting results:

> Although 99% of our eggcases recorded in 2022 came from around the British Isles, we had eggcases recorded from many different countries including: Australia, Canada, Denmark, France, Israel, South Africa and the USA. We love getting photos of eggcases from all around the world, so please keep them coming! Some notable finds include: our first ever record of a Puffadder Shyshark from South Africa; our best year so far for Brown Skate (22 eggcases) and Port Jackson Sharks (23 eggcases); our second ever Chain Catshark eggcase and our third California Horn Shark eggcase, reported from the USA. Stay tuned for the release of our Great Eggcase Hunt Australia identification resources, taking the Great Eggcase Hunt officially into the southern hemisphere.[52]

Ocean Conservancy, founded in 1972 is based in Washington DC. 'We work across the world to ensure a healthy ocean and protect the wildlife and communities that depend on it. The ocean is the responsibility of us all.'[53] Among its major programs: Advancing Ocean Justice; Trash Free Seas; Confronting Climate Change; Protecting the Arctic; Sustainable Fisheries Government Relations ('We're working to ensure that our ocean gets the funding and attention it requires').[54]

The Zurich-based *Shark Foundation/Hai-Stiftung*, founded in 1997, is equally active, its guiding principle being 'Sharks are not threatening us, we are threatening them!'[55] Its website has a 'live' shark death counter, tallying the annual rate at about three per second. The foundation supports and funds many international shark conservation projects and is the Swiss member of the European Elasmobranch Association. The Foundation is active in 26 countries, maintains travelling exhibitions and scientific publications and works on multiple medium- and long-term projects, some being: blue shark populations; the shark trade in India; fisheries in Ghana and Angola; migrations of high sea shark species; whale sharks globally; shark nurseries. Some completed projects: Sixgill sharks in the Mediterranean; shark trade in Greece; nurse sharks; angel sharks; basking sharks (England).[56]

Bite-Back Shark & Marine Conservation is a United Kingdom-based conservation group dedicated to sharks. Formed in 2002, its primary goal is of reducing consumer demand for shark products: 'We defend the high seas on the high street.'[57] And: 'If you want to come face to face with the ocean's most deadly predator, you only have to look in the mirror.'[58] Its campaigns focus on restaurants, fishmongers, supermarkets and retailers. The website lists some recent UK successes as being: 580 health food stores no longer sell shark cartilage capsules; 82 per cent drop in the number of restaurants serving shark fin soup; 115 thousand fewer portions of mako, thresher and blue shark sold each year. This organisation is also taking good aim at a particularly negative and destructive media habit: 'Stop sensational shark reporting!'[59]

New York-based *Shark Angels* is a relatively young organization formed in 2007 with a powerful message:

> Since 2007, members of Shark Angels around the globe have come together to advocate for sharks and our beautiful blue planet by turning fear of sharks into fascination, by empowering the public through education and science, and connecting the passionate to spark meaningful change. We do this through science, education, diving, legislation, and outreach. Through our partnerships with leading international shark scientists, Shark Angels connects the public to ground breaking shark research via innovative technologies to determine where and how to protect sharks. We provide free educational presentations to schools worldwide, thereby creating the next generation of eco-warriors. Our work with other nonprofits and conservation organizations advances the fight for shark fin trade bans and the enactment of laws prohibiting shark finning. And finally, we support responsible shark diving operations to encourage everyone to have the chance to come face-to-face with these fascinating creatures in the wild. Through these efforts, Shark Angels spearheads the fight for one collective goal: to save sharks.[60]

The *Marine Conservation Society* is a community of ocean lovers working around the UK to protect the seas, ' . . . fighting for a cleaner, better protected, healthier ocean . . . we fight for the future of our ocean through people-powered action, with science on our side.[61] Founded in 1983 by David Bellamy and Bernard Eaton, it raises funds through membership, donations and conservation projects. King Charles, when Prince of Wales was President of the Society, and in a speech commemorating the society's twenty-fifth anniversary he noted that among its environmental successes it had been instrumental in securing protected status for the basking shark. Also in his speech he described the seas around the United Kingdom as being 'overfished and awash with rubbish'.[62]

Oceana, founded in 2001 and headquartered in Washington, DC, is dedicated to all forms of marine conservation. Its methodology:

> No organization was working exclusively to protect and restore the oceans on a global scale. To fill the gap, our founders created Oceana: an international organization focused solely on oceans, dedicated to achieving measurable change by conducting specific, science-based policy campaigns with fixed deadlines and articulated goals. Since its founding, Oceana has won more than 275 victories and protected nearly 4 million square miles of ocean. Our science-based campaigns have specific goals and deadlines — enabling us to create real, measurable change for the oceans. From stopping bottom trawling in sensitive habitat areas to protecting sea turtles from commercial fishing gear, Oceana has been a leader in protecting and restoring marine biodiversity and abundance.[63]

The Institute for Marine and Antarctic Studies (IMAS), University of Tasmania. As the name suggests, this Institute operates as a centre of academic excellence in research, focusing on big industries and issues: fisheries and aquaculture; ecology and biodiversity; oceans and cryosphere; climate change; Ocean-Earth systems; Oceans and Antarctic governance.[64] One 2023 IMAS project, modest in scope and funding and restricted to a single remote locality, yet symbolises the threat to elasmobranchs everywhere. This is because the project involves a small little-known, recently discovered skate that has suddenly become critically endangered.

University of Tasmania postgraduate student Erin Woodward received a 2023 Governor's Environment Scholarship to investigate heavy metal bioaccumulation in the soft tissues of the endangered Maugean skate (*Zearaja maugeana*) (Plate 17) in Macquarie Harbour on the west coast of Tasmania. One of the most significant changes to the Harbour's environment was the introduction of large quantities of heavy metal resulting from historical mining. Woodward's project, supervised by Associate Professor Jayson Semmens, head of the skate research program at IMAS, was undertaken through non-lethal and minimally invasive sampling of muscle tissue from a small number of skate (they are mostly confined to the Harbour's shallow upper reaches). Inorganic pollutants such as heavy metals have been linked to stress, reproductive impairment and a variety of health issues across multiple species. This is of particular concern for long-lived higher-trophic species such as sharks and rays, as they will more readily accumulate and magnify these substances to potentially dangerous concentrations. Woodward's project benefits extended beyond helping the conservation of the Maugean skate, in determining how metal accumulations impact other vulnerable sharks and rays in coastal habitats.

The Maugean skate is one of very few known to inhabit

fresh water. The species was first discovered in 1988. Divers report rare sightings, usually in shallow, brackish water, of the animal partly concealed in the sand. This skate shares some close morphological characteristics with western Tasmania's deepwater continental slope skates, rather than with the region's inshore marine skates. This may relate to its feeding habits. Buttongrass tannins stain and darken the waters of Macquarie Harbour, resulting in a lack of light and silty floor, a similar environment to that at continental slope depth. This means that there are no algal communities such as normally grow in light-penetrated water; in their place are invertebrate communities, not dissimilar to those found in the dark depths of the continental slope. And like its deepwater relatives, the Maugean skate may use its elongated, electroreceptive snout to search for food in the silty substrate. Frills at the edge of a prominent nasal curtain may also be sensitive to prey. The adult is quite small, averaging about 70 centimetres in length. Unique as a shallow brackish water specialist, it has an ancient, conservative lineage that may date back to the Cretaceous Period that ended 65 million years ago, when modern geological Tasmania was forming. Yet in recent years its future has become greatly imperilled:

> Scientists have warned an ancient fish species that has survived since the time of the dinosaurs could be one extreme weather event from extinction after its population crashed by nearly half between 2014 and 2021. The Maugean skate, described by a prominent Tasmanian marine scientist as a 'thylacine of the sea', is found only in Macquarie Harbour, a vast body of water on Tasmania's remote west coast. The sharp decline over the past decade

is because its environment has been degraded by human influence, including from salmon farming pollution, hydro power stations altering upstream river flows and rising harbour temperatures due to the climate crisis. These factors have contributed to a sharp drop in dissolved oxygen levels, which particularly affect survival rates in young skates.

The decline is believed to have been hastened by an extreme event in 2019 – an 'inversion' in which the water column was overturned by a westerly storm, lifting poorly oxygenated water from the harbour floor nearer to the surface. Statistically widespread mortalities were found among electronically tracked skates after the event, 11 of 25 tagged individuals dying suddenly.

The health of Macquarie Harbour, an area about six times larger than Sydney Harbour, plummeted last decade after a significant expansion of salmon farming was approved in 2012 despite warnings of potential impact on marine life. The cap on fish farm production was later lowered back to near pre-expansion levels. The health of some species improved, but oxygen levels did not rebound.

A further factor causing alarm is the apparent decline in the number of young skate in the population. Research suggests this could be due to eggs on the harbour floor being exposed to unfavourable oxygen levels.[65]

While it can briefly tolerate warmer surface or saline bottom waters, the species prefers zones between 7.5 and 12.5 metres deep with dissolved oxygen levels of between 60 and 80 per cent. Those conditions are becoming far less frequent. If the skate seeks out oxygen-rich shallower water, it is at greater risk of being snared in gillnets set in some

> areas by recreational fishers to catch flounder and escaped salmon. Restrictions on gillnetting were introduced in 2015 and tightened in late 2022 to reduce accidental by-catch of skate. But restoring oxygen levels remains the biggest challenge.[66]

Researcher Erin Woodward's work on the skate is most important. Her prognosis about its future is not optimistic:

> With the Maugean Skate population being estimated to contain less than 1000 individuals, this species could potentially become the first modern-day elasmobranch extinction within the next couple of decades; a title that I am sure the Tasmanian and Australian Governments do not want on their shoulders.[67]

On 7 September 2023 – National Threatened Species Day, that date marking the death of the last known thylacine in 1936 – Australia's Federal Environment Minister Tanya Plibersek announced funding of over $2.1 million towards saving the Maugean skate, including support for an IMAS captive breeding program.

8

SHARKS AND CREATIVITY

Visions of Hunter and Hunted

> *I have a great picture of New York City and* Jaws *in the summer of 1975 and there are lines all around the block [of the cinema] and I have the same picture in the same theatre at Christmas in the snow and there is snow all over New York and* Jaws *is still on the marquee; there is not a line any more, but the film is still playing.*[1]
>
> Steven Spielberg

The creative depiction of sharks, ancient and modern, is limited only by the artist's imagination and materials to hand. The many attributes we associate with or bestow upon sharks—fear, respect, derision, power, spirituality—have always found expression in painting, sculpture, literature, film, cartoons, craftwork and more besides.

Earlier in this book there was reference to the Australian Dreamtime, and how ancestral spirits created the landscape, one such being Mäṉa the whaler (bull) shark. Similarly, an ancestral sawfish used its powerful toothed rostrum to carve out the Angurugu River on Groote Eylandt in the Gulf of Carpentaria. A modern painting of the event, by the prominent young Aboriginal artist Nekingaba Maminyamanja, includes three creator stingrays (see plate 19). In keeping with the utilitarian philosophies of the Dreamtime, these creator stingrays ensure the wellbeing of people: a man waits to spear them in order to eat.

The three-dimensionality and physical assertiveness of sculpture makes it a good medium to depict the relationship between humans and other life forms, particularly when some form or manifestation of power is intended. And sharks are powerful. A 1.6-metre wood and paint sculpture of the shark-king Gbehanzin (he who took on the might of colonial France) depicts 'the metamorpohosis of a mortal into the awesome being of a Dahomean king', although despite Gbehanzin's 'being armed with foresight and protective imagery, the shifting tide of history proved too powerful to overcome, and thus the subject of this divinatory likeness stands as a defeated figure'.[2]

One of the more curious uses of elasmobranchs for creative expression, which also relates directly to the world beyond the known, is of European origin and was practised at the cusp of the Scientific Revolution, when early ocean-going voyagers returned with their tales of hideous deep-sea monsters. Mischievously and ingeniously, dried skates and rays were cut and otherwise shaped into fantastical-looking apparitions. (Drawings, etchings and paintings of such creatures abound.) Their ventral mouths and gills become spooky faces, their claspers and tails an assortment of legs. A batoid so sculpted was known as a Jenny Haniver. 'Sculpted' is perhaps a generous term, since the aim of the 'sculptor' was often to deceive, as with the example:

> Called both a sea eagle and a flying fish, a 'Jenny Haniver' [was] a forgery made by mutilating a ray to resemble a winged sea monster with a human head. The trick worked, and Ambroise Paré recounted a second-hand tale of how a live specimen was presented to the lords of the city of Quioze. The origin of the name 'Jenny Haniver' is unknown, but the first known illustration of one dates from the 16th century.[3]

Many of these Jenny Hanivers were grotesque, but some of the 'faces' have a certain pathos, a weird beauty which may have convinced some that they were not monsters but mermaids. Their oddity is striking even now, in a world accustomed to more and more sophisticated illusions.

An earlier chapter examined how the symbolic power of sharks was frequently appropriated by indigenous peoples in social, spiritual and political contexts. It has also been shown that European colonisers made a point of attempting to suppress such un-Christian vulgarities as shark worship. Yet modern western culture has shown itself capable of being absorbed by the threatening otherness of sharks, to the extent of creatively influencing us, in complex and surprising ways. By way of illustration: what do sharks and flowers have in common? Very little—with one standout exception. At the beginning of the nineteenth century French Emperor Napoleon Bonaparte, restoring the nation traumatised by its 1789 revolution, turned his attention to Lyons which had been ransacked, the city having been the major producer of fine embroidered silk attire for the aristocracy. Its strong tradition of textile manufacturing and design resumed when Napoleon ordered the re-opening of its Ecole des Beaux-Arts. Appointed as professor of flower painting was a Lyons native,

Antoine Berjon, returned from Paris where he had been an impoverished still life and portrait painter. Now financially secure, his artistic career flourished.

Botanically accurate flower paintings were very popular, closely associated with Romanticism and the wonder of nature. Other natural forms, such as fruit and birds, became part of these aesthetically serene compositions. Berjon, vigorously competing with other artists in the genre, conformed—until in 1819, for reasons unknown, he painted a masterpiece, a subversive work entitled *Still Life with Flowers, Shells, a Shark's Head, and Petrifications* (oil on canvas, 107.7 × 87.4 centimetres, Philadelphia Museum of Art) (Plate 20). Berjon, son of a butcher, beautifully created this:

> A mahogany work table, its drawers pulled open, supports three humble, earthernware containers. The foremost, perilously balanced, on the edge, contains branches of fresh-cut pink roses, droplets of dew still on their petals. A paper broadsheet, its print bleeding through to the verso, has been pinned around them to protect against the thorns. Behind, a brown and yellow vase contains a motley array of peonies, tulips, crab apple blossoms and—jarringly at odds with this sumptuous gathering—a coarse, dry sunflower. Jonquils, narcissus, a columbine, and a tulip lie in the foreground, some falling into the open drawer. On the right, a small, squat jug, visible only in silhouette, holds blue asters and a faintly indicated iris.
>
> To the left, in a strong raking light against the nearly black background, is a pile of shells and sea objects: a branch of coral, a large clam, a conch, and—most mysteriously—the severed head of a shark. The flaps of its roughly cut skin fold behind the conch, its inward-grasping teeth the more

> exposed by the semi-decomposed flesh of the skull. It is a remarkable grouping, sensuous and fresh, sinister and disturbing . . . The brutal intrusion of the shark's head adds a profound sense of disquiet and disarrangement . . . it is, all in all, a tour de force of intense theatricality, certainly meant both to please and repel, seduce and frighten.[4]

This painting may have been Berjon's take on the concept of *vanitas,* the fleeting nature of life, classically depicted as a human skull on top of a pile of books, or he may have been drawn to the physical similarities between profoundly different life forms—rose thorns and shark teeth: 'The introduction of an horrific element into an idyllic setting . . . suggests something of the heightened emotional state that permeates all of the Romantic movement.'[5]

Two famous and intriguing artworks that are creative responses to shark attacks were painted one hundred years apart. John Singleton Copley's *Watson and the Shark* (1778, oil on canvas, 182.1 × 229.7 centimetres, National Gallery of Art, Washington) (Plate 21) and Winslow Homer's *The Gulf Stream* (1899, oil on canvas, 71.4 × 124.8 centimetres, Metropolitan Museum of Art, New York) (Plate 22) are both concerned with complex political issues that were gripping US society at the time and they use the threat and fear of the shark as a potent metaphor.

At first sight, *Watson and the Shark* is a dynamic, tension-filled story. And it really happened:

> Brook Watson loved to swim. He was only acting on a natural impulse, then, when he dove into the green waters of Havana Harbor while on a voyage to the West Indies in 1749. A shark sharing those same waters acted on its own

> natural impulse. Three times it attacked the fourteen-year-old boy. In the first attack, the flesh was stripped from the calf of Watson's right leg. After the second attack, his foot was gone. As the shark rose for its third strike, it was at last driven off by a group of sailors, and Watson was saved.[6]

Yet the painting is significantly more complex, and controversial, than that. The hero—wealthy Watson, who commissioned the painting—is in the water with the beastly shark, forming the base of a compositional pyramid at the apex of which stands a black man, a slave. Copley used classical compositional techniques as a matter of course, the primary convention of which for centuries had been that nobility occupied the apex, animals and slaves the base. Why, with respect to the black man, the deliberate inversion?

Copley didn't say and art historians still cannot agree with one another. A starting point in considering the mystery is that Copley, who had been the pre-eminent portrait painter in Boston, actively sought to become a history painter after he had moved permanently to Europe. History painting was highly regarded, because it did not merely depict something, but used it to contemplate the really big picture: God, man and nature. Another starting point is that polarising social and political divides were hard at work on both sides of the Atlantic, with the Declaration of Independence in 1776 setting British Whig against Tory, New World loyalist against patriot. Furthermore, early calls for the abolition of slavery were considered a great threat to the economic fabric of both Great Britain (by far the greatest slave trader) and her American colonies, so lucrative was the triangular trade that saw rum and other goods leave Boston (and other ports) to be traded for slaves in Africa who were then sold in the Caribbean for sugar and molasses to make the rum back in Britain and her New World colonies.

Watson himself spent many years as a Boston-based entrepreneur engaging in trading with the West Indies. He may or may not have indulged in 'blackbirding', illegal slave trading, undercutting the British monopoly. Slavery was a greatly emotive issue and cut across party lines—his attitude towards it is not known with any certainty. He eventually moved to London where he flourished as a wealthy merchant and became an active conservative, and was Lord Mayor of London for a period. What is also not known is the extent to which he directed Copley, also a conservative and a friend of his, as to how to represent the incident. It has been suggested that the contact in the painting between Watson and the black slave—arms reaching out, rescue rope limp in the black man's hand and around Watson's arm—represents some form of contrition on Watson's part, for profiting at the expense of slavery. It may also be a critique of the hypocrisy of the loyalists wanting freedom from the mother country while enslaving others. Some say the boat and those in it represent the brave new world, the shark voracious Great Britain. For others, Watson's actual shark experience in the New World is symbolic of its danger (revolution) and its 'idle blacks' (a contemporary phrase used to describe the slave's curious posture in a rocking boat of frantic activity). Either way, the shark doesn't come out if it very well; indeed, religious interpretations of the painting juxtapose a chaste, pure white victim imperiled by evil.

At the end of the following century when highly respected artist Winslow Homer painted *The Gulf Stream,* slavery had long since been abolished, the Civil War had been over for

30 years, and yet race relations in the US were at their lowest level, through a grim combination of social Darwinism, institutionalised racism, lynching and conservative theorising that the American Negro race was naturally headed for extinction. Homer came from a line of merchants active in trade with the West Indies and he developed a lifelong interest in that part of the world and, as a result, in its peoples. According to one interpretation, Winslow Homer

> . . . clearly assimilated every iconographical ingredient of Copley's *Watson* in *The Gulf Stream.* Yet in stripping away every figure except the black man—thereby making him the central focus of the triangular design—Homer constructed a new meaning based on his understanding of contemporary race relations . . . If the black man in Copley's picture embodies an intellectual abstraction of the libertarian viewpoint, Homer's besieged fisherman is an allegory of the black man's victimization at the end of the nineteenth century.[7]

Thus the four visible sharks around the boat, together with the large pieces of sugar cane, recall a grim story of slavery: thousands of sick Africans were thrown overboard to sharks on the trans-Atlantic crossing (partly for the ship owners to claim insurance). Sometimes slaves were even used as bait to catch sharks to feed to other slaves. Could their lot have got any worse? Yes. If the work seems heartless, it is intended to be. Sharks, a broken boat, a fierce waterspout bearing down and a ship that won't be offering any assistance: it's overwhelmingly suggestive of the man's extinction by both the force of nature and a superior breed of human being. By this interpretation the unrelieved pessimism of *The Gulf Stream*

> ... is a response to the predatory racism that gripped the nation at the turn of the century ... any liberal Republican vision there might have been of achieving social justice genteelly was replaced by withdrawal and even cynicism ... there was little for white liberals to do but what viewers of The Gulf Stream do: look on, aghast, at the degradation of a race that was now fatally surrounded by sharks.[8]

The most controversial contemporary artistic treatment of a shark is British artist Damien Hirst's 1991 sculpture *The Physical Impossibility of Death in the Mind of Someone Living,* which *The Guardian* newspaper calls 'one of the most famous icons of modern art'[9], but which has been controversial since first commissioned for £50 000 by advertising guru Charles Saatchi. Its technical description: 'Tiger shark, glass, steel, 5 per cent formaldehyde solution, 213 × 518 × 213 cm'. The 4.27-metre tiger shark was sourced from Australia at a cost of £6000. The container, classically known as a vitrine, a glassed-in showcase, was made by aquarium specialists, being three cubes bolted together. Late in 2004 the US hedge fund manager Steve Cohen bought the artwork for a reputed £6.5 million, setting up Hirst to become the world's most expensive living artist.

Hirst had not, however, adequately preserved the shark's deep tissue and it began to decay and change shape, while the water grew murkier and murkier. A taxidermy solution didn't work, and he therefore replaced it with a large female tiger shark sourced from Queensland, described as a female of between 25 and 30 years old and weighing 1.92 metric tonnes. The shark was to have originally been a great white but apparently Australian legislative protection of that species came into force a few days before he was to place his order. All of that aside, what is the meaning of this sculpture nicknamed the pickled shark—the single ever most valuable, stared at and discussed dead animal?

Hirst is preoccupied with 'the ambiguity of human experience'[10] and his art is intended to reflect this ambiguity. He said about the work, 'The sculpture brings hunter and hunted face to face. Why are we so fascinated by animals that kill and eat us?'[11] The second sentence of the statement may be true enough; the first only correct when viewing the shark not as the hunter but the hunted. Sharks, for this artist, 'play havoc with our value systems, they make us aware that we are meat, part of the food chain'.12 It is an interesting artistic echo of what has gone before: in Winslow Homer's time the laws of thermodynamics were in vogue and applied beyond physics and thus it was that the black man on the broken boat eating his sugar cane, then being eaten by the shark, merely represented part of a transfer-of-energy chain, with all matter eventually and equally subject to entropy. And in a connection to both *The Gulf Stream* and *Watson and the Shark, The Physical Impossibility of Death in the Mind of Someone Living* is also unresolved. 'The work offers drama without catharsis, confrontation without resolution, and provocation without redress. Responsibility is returned to the viewer.'[13] It is a safe bet that neither of Damien Hirst's tiger sharks tasted human flesh.

Hirst's fascination with bodies and the decomposition of bodies is part of a long tradition. Artistic absorption with mortality is as old as civilisation. As noted earlier, perhaps the earliest known depiction of a shark attack appears on a 725 BC piece of ancient Italian pottery. Two millennia later, Damien Hirst described photographs of human (war) wounds as 'completely delicious, desirable images of completely undesirable and unacceptable things'.[14] In the broad context of physical human suffering, it is hard to equate the fear of a shark bite with art; perhaps, in the eternally dead tiger shark in Hirst's tank, there is a soundless echo of our dark impulses. Not surprisingly, Hirst the artist has been attacked, not by sharks but animal rights campaigners.

Hirst's work inspired the innovative high-profile Czech sculptor David Cerny to continue the tradition of shark art as enabling an implicitly potent political message. (Cerny's works include his *Pink Tank,* in which, one night, he painted pink a Soviet tank mounted on a huge plinth in Prague, commemorating the 1945 liberation of Poland from the Germans by Soviet forces.) To make a statement about the US-led invasion of Iraq in 2003, Cerny created a sculpture entitled *Shark:* 'It features a life-size Saddam Hussein in underpants with his hands tied behind his back, floating in a large glass tank filled with the embalming fluid formaldehyde.'[15] Interestingly, the sculpture was banned from its proposed display in a Belgian town in the wake of the violence caused by Danish cartoons of the Prophet Mohammed.

Shark, *2007, by David Cerny.* (David Cerny)

Less controversially, but asking similar questions of the human attitude towards 'dangerous' animals, British environment artists Olly and Suzi's 1997 *Shark Bite* is a crude drawing, in blood and acrylic on paper, of a shark. The work was floated in water off the South African coast and a great white shark photographed biting it—the photo being part of the work, as was the recovered but torn original work. (Their animal interaction doesn't always go according to plan, with instances of 'a leopard dragging a painting away and destroying it, and a rhinoceros eating a whole piece'.[16])

The Headington shark, by John Buckley. (Stephanie Jenkins)

Still in Britain—a country identified with eccentricity—the town of Headington near Oxford, many miles inland, has since 1986 had the most unlikely tourist attraction: an eight-metre fibreglass great white shark impaled in the roof of a suburban terraced house. This wacky sculpture has, in fact, a deadly serious reason for its existence—one that can be said to be avowedly political, while also concerned with the well-being of the planet. The work of art, called *Untitled 1986,* by English artist John Buckley, was commissioned by the owner of the house, Bill Heine, a cinema owner and Oxford radio presenter. Heine had it erected on the forty-first anniversary of the atomic bombing of Nagasaki, explaining that 'the shark was to express someone feeling totally impotent and ripping a hole in their roof out of a sense of impotence and anger and desperation . . . It is saying something about CND, nuclear power, Chernobyl and Nagasaki.'[17] In keeping with such an unlikely use of a shark—although its massive 'alien' power is no doubt part of the statement—the sculpture has its own all-too-human story:

> It had been winched up by a crane overnight, and although the police were aware of what was going on they were powerless to do anything, as there is no law to prevent a man from putting a shark on his own roof . . . Oxford City Council tried to get rid of the shark on the grounds that it was dangerous to the public, but engineers inspected the roof girders that had been specially installed to support it and pronounced the erection safe. The council then decided that the shark was a development . . . and that as such it had to be removed. Their offer to display it in a public building such as a swimming pool was not, however, accepted by Bill . . . in 1990 he was refused retrospective planning permission by Oxford City Council. Undeterred, in 1991 he appealed to the Secretary of State for the Environment whose office came out in favour of the applicant, stating: It is not in dispute that this is a large and prominent feature.

> That was the intention, but the intention of the appellant and the artist is not an issue as far as planning permission is concerned . . . it is not in dispute that the shark is not in harmony with its surroundings, but then it is not intended to be in harmony with them. The basic facts are there for almost all to see. Into this archetypal urban setting crashes (almost literally) the shark. The contrast is deliberate . . . and, in this sense, the work is quite specific to its setting . . . The Council is understandably concerned about precedent here [but] . . . any system of control must make some small place for the dynamic, the unexpected, the downright quirky. I therefore recommend that the Headington shark be allowed to remain.'[18]

A shark in the roof of the house is one thing; a house that *is* a shark is something else altogether. Mexican architect Javier Senosiain, born in 1948 and taking inspiration from both Frank Lloyd Wright and Antoni Gaudi, specialises in organic domestic architecture. He lives in one of his more celebrated creations in Mexico City (that city founded upon Cipactli's elasmobranch-related sacrifices):

> Senosiain's home, which he shares with his wife, Paloma, and their daughters, looks like an enormous shark set into a hillside—the dorsal fin protruding from the roof eliminates any doubt . . . In the shark's gaping jaws, the curved window of Senosiain's upstairs studio overlooks the city. Another studio window, a small porthole, forms the shark's eye . . . He finished the 1,800-square-foot interior in a smooth stucco made of white cement, beige mortar, and marble dust. For the floor, he picked beige wall-to-wall, inviting that intimate contact between flooring and

bare feet. To reinforce the connection between man and his most elemental shelter, shoes must be removed upon entry. The ground floor of the shark functions primarily as a passageway. From here, a staircase ascends to the second floor—a womblike telephone niche built into the underside of the stuccoclad structure . . . Beyond the staircase, the ground floor of the shark offers access to a tunnel. At the far end of this mysterious passage—illuminated by an unseen skylight—bedrooms and living spaces are half-buried underground . . . In the master bedroom, cartoonlike cutouts framing closet niches impart a Flintstones air. A leather-cushioned cast-in-place ferrocement bench spirals outward, like a nautilus shell, and turns into the bed, nestled in a curving wall. 'The idea is to recline,' says Senosiain, 'like an animal in a cave.'[19]

Javier Senosiain's shark house. (Francisco Lubbert)

Sharks have their inevitable place in creative literature, as irredeemable adversaries to mankind. Five examples follow, each mirroring their time. The first is a breathtaking illustration of how popular literature (that is, read uncritically by lots of people) can capture and infect the popular imagination, in this case casually blending racism with woefully inaccurate depictions of sharks. The author, W.H.G. Kingston, an Englishman who spent most of his life in Portugal, was a prolific and successful author of tales for boys. His 1876 novel, *The Three Lieutenants,* includes this gem of a bad-taste scene set in the harbour of Kingston, Jamaica:

> Jack and Terence sailed up to Kingston with a fresh sea breeze a-blowing over the sandy shore of the Palisades.
>
> 'Take care you don't capsize us,' said Jack to the black skipper, who carried on till the boat's gunwhale was almost under water.
>
> 'Neber tink I do dat, massa leetenant. Not pleasant place to take swim,' answered the man, with a broad grin on his ebony features, showing his white teeth.
>
> 'I think not, indeed,' exclaimed Terence. 'Look there.'
>
> He pointed to a huge shark, its triangular fin just above the surface, keeping two or three fathoms off, even with the boat, at which the monster every now and then, as he declared, gave a wicked leer.
>
> 'What do you call that fellow?'
>
> 'Dat, massa, dat is Port Royal Jack,' answered the negro. 'He keep watch ober de harbour—case buckra sailors swim ashore. He no come up much fader when he find out we boat from de shore. See he go away now.'
>
> The shark gave a whisk with his tail, and disappeared in an instant. The young officers breathed more freely when their ill-omened companion had gone. Almost immediately afterwards a boat belonging to a large merchantman, lying

at the mouth of the harbour, ready for sea, passed them under all sail. Her crew of eight hands had evidently taken a parting glass with their friends.

'Dey carry too much canvas wid de grog dey hab aboard,' observed the black. 'Better look out for squalls.'

He hailed, but received only a taunting jeer in return, and the boat sped on her course. Not a minute had passed when Jack and Terence heard the negro mate, who was watching the boat, sing out-

'Dere dey go, Jack shark get dem now- eh?'

Looking in the direction the black's chin was pointing, to their horror they saw that the boat had capsized, her masts and sails appearing for an instant as she rapidly went to the bottom, while the people were writhing and struggling on the surface, shrieking out loudly for help. Jack and Terence ordered the black to put the boat about instantly, and go to their rescue. Nearly two minutes passed before they reached the spot. Five men only were floating. The ensanguined hue of the water told too plainly what had been the fate of the others . . . Influenced by a generous impulse, and forgetting the fearful monsters in the neighbourhood, [Jack] was on the point of leaping overboard, when the black boatman seized his arm, crying out,-

'No, no, massa, dat one shark, hisself.'

Jack looked again, and the object he had mistaken for a seaman's white shirt resolved itself into the white belly of a shark, the creature being employed in gnawing the throat of its victim.

'Dat is what dey always do,' observed the black coolly. 'Dey drag down by de feet, and den dey begin to eat at de trote.'[20]

Herman Melville's *Moby-Dick,* first published in 1851, is one of the great works of fiction, a huge and hugely ambitious novel that, in the guise of one man's obsessive hunt for a near-mythical whale, questions the very meaning of existence. The power of the writing is everything that Kingston's is not, not least because Melville relied on relative accuracy (in this case a dramatic feeding frenzy) to set the scenes for his intellectual embellishments. Here, as narrated by the book's protagonist Ishmael, a captured sperm whale has been hauled alongside the whaling vessel *Pequod,* of which mad Ahab is the Master. Normally, the whale would be left lashed alongside the boat overnight:

> But sometimes, especially upon the Line in the Pacific, this plan will not answer at all; because such incalculable hosts of sharks gather round the moored carcase, that were he left so for six hours, say, on a stretch, little more than the skeleton would be visible by morning. In most other parts of the ocean, however, where these fish do not so largely abound, their wondrous voracity can at times be considerably diminished, by vigorously stirring them up with sharp whaling-spades, a procedure notwithstanding, which, in some instances, only seems to tickle them into still greater activity. But it was not thus in the present case with the Pequod's sharks; though, to be sure, any man unaccustomed to such sights, to have looked over her side that night, would have almost thought the whole round sea was one huge cheese, and those sharks the maggots in it.
>
> Nevertheless, upon Stubb setting the anchor-watch after his supper was concluded; and when, accordingly, Queequeg and a forecastle seaman came on deck, no small excitement

> was created among the sharks; for immediately suspending the cutting stages over the side, and lowering three lanterns, so that they cast long gleams of light over the turbid sea, these two mariners, darting their long whaling-spades, kept up an incessant murdering of the sharks, by striking the keen steel deep into their skulls, seemingly their only vital part. But in the foamy confusion of their mixed and struggling hosts, the marksmen could not always hit their mark; and this brought about new revelations of the incredible ferocity of the foe. They viciously snapped, not only at each other's disembowelments, but like flexible bows, bent round, and bit their own; till those entrails seemed swallowed over and over again by the same mouth, to be oppositely voided by the gaping wound. Nor was this all. It was unsafe to meddle with the corpses and ghosts of these creatures. A sort of generic or Pantheistic vitality seemed to lurk in their very joints and bones, after what might be called the individual life had departed. Killed and hoisted on deck for the sake of his skin, one of these sharks almost took poor Queequeg's hand off, when he tried to shut down the dead lid of the murderous jaw.[21]

Edgar Allan Poe, whose legacy to world literature includes his prominence as an early modern short story writer and his invention of the modern detective story, was frequently drawn to the macabre. Certainly, he wrote suffering cloyingly well, as in this account of shipwreck survivors in his only novel, *The Narrative of A. Gordon Pym,* published in 1838:

> We now saw clearly that Augustus could not be saved; that he was evidently dying. We could do nothing to relieve his sufferings, which appeared to be great. About twelve

o'clock he expired in strong convulsions, and without having spoken for several days. His death filled us with the most gloomy forebodings, and had so great an effect upon our spirits that we sat motionless by the corpse during the whole day, and never addressed each other except in a whisper. It was not until some time after dark that we took courage to get up and throw the body overboard. It was then loathsome beyond expression, and so far decayed that, as Peters attempted to lift it, and entire leg came off in his grasp. As the mass of putrefaction slipped over the vessel's side into the water, the glare of phosphoric light with which it was surrounded plainly discovered to us seven or eight large sharks, the clashing of whose horrible teeth, as their prey was torn to pieces among them, might have been heard at the distance of a mile. We shrunk within ourselves in the extremity of horror at the sound . . . During the whole day we anxiously sought an opportunity of bathing, but to no purpose; for the hulk was now entirely besieged on all sides with sharks—no doubt the identical monsters who had devoured our poor companion on the evening before, and who were in momentary expectation of another similar feast. This circumstance occasioned us the most bitter regret and filled us with the most depressing and melancholy forebodings. We had experienced indescribable relief in bathing, and to have this resource cut off in so frightful a manner was more than we could bear. Nor, indeed, were we altogether free from the apprehension of immediate danger, for the least slip or false movement would have thrown us at once within reach of those voracious fish, who frequently thrust themselves directly upon us, swimming up to leeward. No shouts or exertions on our part seemed to alarm them. Even when one of the largest was struck with an axe by Peters and much

> wounded, he persisted in his attempts to push in where we were. A cloud came up at dusk, but, to our extreme anguish, passed over without discharging itself. It is quite impossible to conceive our sufferings from thirst at this period. We passed a sleepless night, both on this account and through dread of the sharks.[22]

It is generally believed that Ernest Hemingway's novella *The Old Man and the Sea,* published in 1952 and the last of his works published before his self-inflicted death, was instrumental in his being awarded the Nobel Prize for Literature. The storyline is simple. Santiago, an ageing Cuban fisherman, after a long period of poor luck, hooks a huge 5.5-metre marlin while fishing in his skiff, far out in the Gulf Stream. He then has to contend with sharks. Hemingway, a noted big game fisherman, wrote the story as an allegory of the human condition, with religious overtones—Santiago being a Christ-like individual, suffering the 'natural' cruelty of defeat hard in the wake of triumph (a complexity perhaps matching the author's then state of mind).

Given the single-minded power of this work of fiction, and its place in the canon of western literature, it may be considered regrettable to reduce it here to a *Reader's Digest*-like mauling of a few hundred words. But this is, after all, only a book about sharks, and Hemingway knew a few in his time:

> He was a very big Mako shark, built to swim as fast as the fastest fish in the sea and everything about him was beautiful except his jaws . . . all of his eight rows of teeth were slanted inwards. They were not the ordinary pyramid-shaped teeth of most sharks. They were shaped like a man's fingers when

they are crisped like claws . . . The shark closed fast astern and when he hit the fish the old man saw his mouth open and his strange eyes and the clicking chop of the teeth as he drove forward in the meat just above the tail. The shark's head was out of the water and his back was coming out and the old man could hear the noise of skin and flesh ripping on the big fish when he rammed the harpoon down . . . the old man hit it. He hit it with his blood-mushed hands driving a good harpoon with all his strength. He hit it without hope but with resolution and complete malignancy . . . He could see their wide, flattened shovel-pointed heads now and their white tipped wide pectoral fins. They were hateful sharks, bad smelling, scavengers as well as killers, and when they were hungry they would bite at an oar or the rudder of a boat . . .

They came in a pack and he could only see the lines in the water that their fins made and their phosphorescence as they threw themselves on the fish. He clubbed at their heads and heard the jaws chop and the shaking of the skid as they took hold below. He clubbed desperately at what he could only feel and hear and he felt something seize the club and it was gone . . . When he sailed into the little harbor the lights of the Terrace were out and he knew everyone was in bed . . . he shouldered the mast and started to climb. It was then he knew the depth of his tiredness. He stopped for a moment and looked back and saw in the reflection from the street light the great tail of the fish standing up well behind the skiff 's stern. He saw the white naked line of his backbone and the dark mass of the head with the projecting bill and all the nakedness between.[23]

Career journalist Peter Benchley, like countless other young would-be novelists with a dream, earned money to buy writing time: 'I sat in the back room of the Pennington Furnace Supply Co. in Pennington, New Jersey, in the winters, and in a small, old turkey coop in Stonington, Connecticut, in the summers, and wrote what turned out to be *Jaws*.'[24] He had grown up on Nantucket Island, Massachusetts, and developed an interest in sharks. A direct spark for the novel came from his reading an account of a huge great white shark hooked off Long island. So significant was the manuscript of what became the 33-year old's first novel, published in 1974, that 'within eight weeks, it had leaped to No.2 position on the *New York Times* bestseller list. Before the book was even off the presses, it had already earned over $1 million, including $575,000 for US paperback rights alone, and from sales to book clubs, foreign publishers and the film's producers.'[25]

Even without the subsequent film adaptation, it was a significant literary achievement, combining the thoughtful luck of rendering original an old storyline concept—monster terrorises innocents—with a well-paced plotline and above-average writing. Then came the movie, and the novel surfed a gigantic wave, selling over twenty million copies. Benchley's view of the novel: 'Completely inadvertently, it tapped into a very, very deep fear . . . If I had done it on purpose, it would be one thing. But I didn't know for years what was responsible for the enormous phenomenon of *Jaws*.'[26]

The mystery is not entirely solved, except to say that both book and film belong in a fictional genre that has mass appeal, extending as far back as the legend of Jonah and the whale (possibly a whale shark or basking shark) and contemporaneously to such influences as Herman Melville's novel *Moby-Dick* (1851), Ibsen's play *An Enemy of the People*

(1882), the films *The Creature from the Black Lagoon* (1954) and *Psycho* (1960), the *King Kong* movies and the documentary *Blue Water, White Death* (1971). The true tragedies of the New Jersey attacks (1916) and the USS *Indianapolis* shipwreck attacks (1945) were also horror stories waiting to be dramatised.

Sharks were well on the nose in the 1970s: other than as a class of boneless fish to go with chips, they were not economically lucrative, and nothing had happened to make their other known feature—people as prey—attractive. Then came *Jaws.* The potential within the book morphed into a record-breaking Hollywood success story, all based on the myth of a gigantic great white shark turned rogue with a taste for revenge and human flesh. However, unlike other monster stories, it had a basis in reality. Where there could never really be a gigantic ape hanging on to a tall building for all the world to see, there could definitely be something of menace under the waves. This lent to the movie legitimacy and credibility, no matter how unlikely and scientifically inaccurate its premise.

The young director, Steven Spielberg, with just two films to his name, co-wrote a screenplay (constantly changed during the very long shoot) markedly different to Benchley's novel, stripping much of it away for two essentials: the shark, and the relationship between the three main characters. Crucially, Spielberg included a retelling of the USS *Indianapolis* tragedy as a key to the storyline. This use of reality enhanced the legitimacy of *Jaws* and therefore the shark—subsequently almost any shark—as a malevolent, virtually satanic beast.

So powerful was this creature that Spielberg used it creatively and artfully to address some socio-political issues and to pose some interesting psychological questions, in the process making a film that is considered to be a classic. The cinematic storyline: a huge shark starts mauling bathers at Amity Beach, New Jersey. The town relies on tourism. New police chief Brody (Roy Scheider) wants to close the beaches but comes up against the smarmy mayor. After more deaths, Brody enlists the aid of marine biologist Hooper (Richard Dreyfus) and rough, tough shark fisherman Quint (Robert Shaw) to hunt and destroy the shark. They do so, but at the cost of Quint's life.

Jaws neatly inverted what had been a Hollywood tradition of male heroes (created by macho male directors) by killing off Quint, having first set him up as an obsessive with a death-wish endangering the lives of all three men. It is as if an audacious young novelist at the time had written an international bestseller ridiculing the shark-hunting exploits of Ernest Hemingway and Zane Grey. The inversion sees the manly Quint suffer the ultimate fate/indignity of sliding screaming into the great beast's maw. In this respect Spielberg continued the tradition of both Copley and Homer, using a shark to subvert convention.

Quint, though, is also an echo of the complexities inherent in Melville's Captain Ahab. Quint is a USS *Indianapolis* survivor. His monologue to Brody and Hooper about that horror is a powerful piece of cinema and the core of the film. The deaths of the teenagers at the beginning were scary: Quint's monologue is chilling.

Politically, the movie can be read as a kind of allegory of the turbulence and trauma associated with the Vietnam War and the Watergate scandal, in which there were clear demarcations of the left (brainy scientist Hooper) and the right (straight-talking hunter Quint)—with Brody somewhat powerless in the middle, and tainted by the corruption of officialdom that kept the beach open and led to more deaths. How ironic it is that the politics of *that* shark (Spielberg nicknamed it Bruce, after his lawyer) has led directly to all sharks becoming political, as sides line up for an almighty twenty-first century conservation–exploitation conflict.

Noteworthy, too, because it reinforces our uneasy notions about sharks, *Jaws* is both a tremendously tense suspense movie and a less successful action movie. The shark is not seen (dorsal fin excepting) until far into the film. This withholding is most effective—even if it may have been forced on Spielberg through the mechanical shark repeatedly breaking down in water—because in reality that's what we fear most about sharks. We cannot see them coming for us until it is too late. The first full view of the shark is by Brody (the movie's eventual mundane hero) on Quint's boat. It glides right past his stunned face, in a richly complex scene:

> All the things this creature has been! A dirty old man racked with longing [eating beautiful teenagers]. An insatiable psychopath forced to repeat a sin. A scarlet pimpernel leaving a toothy plume. An insolent catwalk model. A Bond adversary salivating at the possibility of an equal opponent. But from this moment he's also simply a species enemy. He now exists as a shark. He exists, and like all monsters, he is *far, far older than us.* And what jerks Brody back, so speedy and rigid it's as if *he* were the special effect, is the shark's silence. Even though its mouth was open there came no growl, no thunder, no list of impossible requests. No proof of this encounter beyond a cool fizz on the water.[27]

In July 2023 a US-made documentary called *Sharksploitation* (directed by Stephen Scarlata, Shudder streaming service) premiered in a number of countries. It is an account of the neverending post-*Jaws* subgenre of shark exploitation movies. Nearly half a century after the release of *Jaws* in 1975, the list of such movies stands at over 70 – that's a lot of material to work with, and

> *Sharksploitation* succeeds as a love letter to one of cinema's most curious subgenres because it considers all shark films to be part of the same family, thus giving its subject the proper documentary treatment . . . examining their impact and cultural significance.[28]

Interviewees include sharksploitation filmmakers, academics and marine biologists. 'Rather than merely dismissing shark movies as nothing more than schlock, the experts seriously consider how shark films speak to us through the medium.'[29] Furthermore,

> As suggested by the interviews, shark movies seem to speak to a collective kind of thalassophobia, or a fear of deep water. The brand of thalassophobia activated by shark films expresses our anxieties about what might be lurking beneath the surface, especially that which cannot be seen . . . By demonstrating their respect for the cinematic power of shark films, the interviewees avoid mocking sharksploitation, even when they explore the most ludicrous of films.[30]

Sharksploitation films from 1975 to 2023 can be grouped in categories, including: Thriller; Adventure; Sci-fi/Horror; Survival/Horror; Creature feature; Horror/Comedy; Action/Adventure/Comedy; Sci-fi/Disaster/Comedy.

Titles include: *Night of the Sharks*; *Deep Blood*; *Cruel Jaws*;

Sharkenstein; *Shark Attack 3: Megalodon*; *Sharknado 5: Global Swarming*; *Open Water*; *Red Water*; *12 Days of Terror*; *Blue Demon*; *Hammerhead: Shark Frenzy*; *Raging Sharks*; *Shark Swarm*; *Sharks in Venice*; *Mega Shark Versus Crocosaurus*; *Dinoshark*; *Sharktopus*; *2-Headed Shark Attack*; *Dark Tide*; *Avalanche Sharks*; *Ghost Shark 2: Urban Jaws*; *Roboshark*; *Sharkansas Women's Prison Massacre*; *47 Meters Down: Uncaged*; *The Reef: Stalked*.

Ludicrous schlock or not, the collective value of these cinematic cultural artefacts, in generating respect and caring for sharks, can be inferred here:

> Peter Benchley (late author the 1974 novel *Jaws*) makes numerous archival appearances throughout *Sharksploitation* demonstrating the film's commitment to including the author as the originator of the subgenre. Some of the most affecting interviews come from Wendy Benchley, his wife, who discusses *Jaws*' impact not just on the world, but also on Benchley himself, who spent many years, alongside Wendy, as a fierce ocean activist.[31]

9

CREATURES OF EXTREMES

Descriptions of Sharks, Skates, Rays and Chimaeras

> *The squaliform sharks are creatures of extremes: in size they range from the puny to the downright gigantic, they inhabit a wide range of depths, from sundappled shallows to the chill blackness of the abyss, and their taxonomy is a veritable morass of contention and tentative revision'.*[1]

This quotation is no exaggeration and it does not always appear logical how and why shark species, genera, families and orders are arranged. This reflects the highly complex and mysterious nature of neoselachian evolution and the more fundamental reality that appearances can be deceptive. The literature does sometimes arrange sharks by categories other than their scientific orders, for example by size, geographic distribution, or whether inshore or pelagic. This extended chapter describes a representative selection of sharks, skates, rays and chimaeras arranged by their scientific orders.

THE SQUALEOMORPHS: DOGFISHES, SAWSHARKS, ANGEL SHARKS, COW SHARKS, FRILLED SHARKS

Squaliformes

This order of sharks, commonly called the squaloids or dogfish sharks, has the following characteristics:

Classification

- Approximately 22 genera in seven families
- Approximately 107 species

Biology

- No nictitating membrane
- Five gill slits
- Two dorsal fins, often with spines
- No anal fin
- Bioluminescence in many species
- Ovoviviparous reproduction

Habitat

- Estuarine, coastal, oceanic
- No freshwater species
- Widespread globally from Arctic to Antarctic waters
- Mostly bottom dwellers
- Nightly vertical migration common in many species

Dwarf lanternshark (*Etmopterus perryi*)

Found in deep Atlantic and Caribbean waters off South America, the adult of this tiny shark reaches a maximum of nineteen centimetres in length. What does such a truly diminutive predator eat? Probably larger krill and shrimps, and it is thought that dwarf lanternsharks feed cooperatively, attacking squid larger than themselves. Discovered in 1985, in the Caribbean Sea, the dwarf lantern shark is currently listed as the world's smallest known shark.

The much larger southern lanternshark (*Etmopterus baxteri*) grows to about 60 centimetres and inhabits cool southern Pacific and Atlantic waters. Lanternsharks congregate over seamounts in waters near Tasmania, and as a result became an orange roughy bycatch, but the soft flesh is not considered commercially viable.

Spined pygmy shark (*Squaliolus laticaudus*)
A widely distributed deepwater tropical pelagic species, the adult of which grows to just 22 centimetres. This shark spends its days at the bottom of the ocean and at night follows its vertically migrating prey—deepsea squid, bristlemouth fish and lanternfish. Bioluminescent organs on the shark's belly attract prey in the lightless depths.

Little is known of its biology.

Prickly dogfish (*Oxynotus bruniensis*) (Plate 1)
Also known as the roughshark, the prickly dogfish inhabits deep southern Australian and New Zealand waters. A small shark growing to at least seventy centimetres, it has dermal denticles that are large and prickly and firm ridges running the length of its belly. Colouration is generally brown to grey. Contrary to the popular image of a shark as sleek and torpedo-shaped, the prickly dogfish is best described as humpbacked or cambered in shape. It also has large floppy dorsal fins, with barely visible spines firmly embedded at their bases.

Not much is known about the biology or behaviour of this species, although there is some speculation about its close relative, *Oxynotus centrina*: 'Angular roughsharks feed upon polychaetes, perhaps utilising their very large livers to hover at neutral buoyancy above the seafloor with limited forward motion whilst seeking prey'.[2] The teeth of *O. bruniensis* are lanceolate in the upper jaw and blade-like in the lower jaw, to maximise its ability to clutch and cut its prey. Interestingly, this tooth structure is similar to that of the cookie-cutter shark.

Cookie-cutter shark (*Isistius brasiliensis*)

Out of all the ways in which one creature can eat another, the cookie-cutter shark has evolved a method that is truly audacious. This tiny shark reaches about 50 centimetres in length when fully grown, but preys upon marine creatures hundreds of times larger than itself. The cookie-cutter shark is a deep-water dweller and its skin is sprinkled with photophores, which make it glow greenly in the dark to attract prey such as squid, but the cookie-cutter's primary weapon is its basihyal. In conjunction with powerful, suction-plated lips, the basihyal vacuum seals the rounded mouth onto the chosen pelagic prey—whale, tuna, marlin, seal, dolphin. The little mouth is full of teeth, of varying shapes, which bite into the victim as the cookie-cutter twists its body around, extracting a round plug of flesh and leaving a wound up to seven centimetres deep. Predator or parasite? Both. Officially, the cookie-cutter shark is a facultative ectoparasite: it exists both as an external parasite and independently of parasitism.

Portuguese shark (*Centroscymnus coelolepis*) (Plate 2)

The body plan and teeth structure of this shark are very similar to those of the Greenland shark, but there the similarity ends, for despite the close taxonomic relationship, the Portuguese shark seldom grows to more than a metre in length, while the Greenland shark may exceed seven metres. The Portuguese shark is a benthic dweller, reflected in its colouring: the pup's skin is almost black, becoming chocolate-brown as it matures. This species was heavily fished off Portuguese waters, hence its common name. It is also found in waters off Newfoundland, Brazil, southern Australia, western Namibia and the western Pacific Ocean. A Portuguese shark has been caught at 3576 metres—the record known depth for a shark. The Portuguese shark's large denticles overlap closely. It seems to have a fairly broad diet.

Whitespotted spurdog (*Squalus acanthias*)
This shark has a variety of names including spiny dogfish and piked dogfish and it also apparently attains a variety of maximum sizes according to habitat, from about one metre in Australian waters to twice that in the Black Sea. Furthermore, it has the longest known gestation period of any shark—up to two years—with a lifespan of about 70 years. It is widespread globally in both inshore and offshore waters, forms large schools, and is highly migratory. Unfortunately for the whitespotted spurdog, it could also be known as the 'fish-andchips' shark. It has been so overfished in the north-eastern Atlantic that it is listed as endangered on the IUCN Red List and strict limits have been set on catches. In Australia it is abundant and does not have much culinary appeal, being considered 'to be rather coarse'.[3] In fact Australians came in for a bit of a scolding in the early 1960s for their unadventurous taste buds:

> On occasions many tons of the Dogfishes have been brought up in a short period. At first practically all had to be thrown back as there was no market for the fish; but gradually the public were educated to buying 'Flake', the

> name under which the flaked Dogfish flesh is well (and so favourably) known in Great Britain, and some soon learned to appreciate it. But most of the Australian people are extremely touchy about trying anything in the fish line that they cannot determine on its appearance; so it is always an uphill fight to introduce them to, and to get them used to buying and using for the table, anything that seems to them to be 'new'.[4]

Kitefin shark (*Dalatias licha*)

This shark is somewhat similar in appearance to the Portuguese shark, but is larger, reaching an average of about 1.4 metres in length. It has a patchy worldwide distribution, with concentrations in the eastern Atlantic and in southern Australian and New Zealand waters. The kitefin shark is found at depths of around 600 metres but has been recorded at depths of nearly 2000 metres. Its colouration is variously described as dark chocolate brown, cinnamon, and violet brown. This is a hovering shark, its large liver enabling it to hang almost motionless over the outer shelves and slopes of its deepwater habitats. Prominent features of this shark are its large eyes and thick lips in which are powerful jaws containing large teeth: the kit of a formidable predator. In turn, it is targeted by larger sharks occupying the same niche. The second dorsal fin is slightly larger than the first. It has one of the largest elasmobranch livers—about one fifth of its body weight.

Greenland shark (*Somniosus microcephalus*)

This shark is popularly described as a sluggish carrion eater (and its scientific name translates as 'sleepy tiny-brain') but the truth is that it is one of the planet's cold saltwater apex predators (another being the orca). The Greenland shark commonly exceeds four metres in length and specimens measuring more than seven metres have been recorded. The Greenland shark weighs up to 1000 kilograms and its primary habitat is the seas and fjord systems of Greenland and the Canadian Arctic, although it is found throughout Arctic waters. Southern populations inhabit Canadian waters including the Gulf of St Lawrence. To the east, they are common in Scandinavian waters. It is the only shark known to live beneath Arctic ice sheets, where it has been recorded in water temperatures of minus 1.94 degrees Celsius. An unmanned submarine recorded a six-metre Greenland shark off the east coast of the United States at a depth of 2200 metres. The Greenland shark has high concentrations of TMAO (trimethylamine oxide), which acts as an anti-freeze and a protein stabiliser at great depths, and as in other shark species is an important component of the osmoregulatory system. This compound is present at neurotoxic concentrations and renders the flesh toxic unless correctly processed.

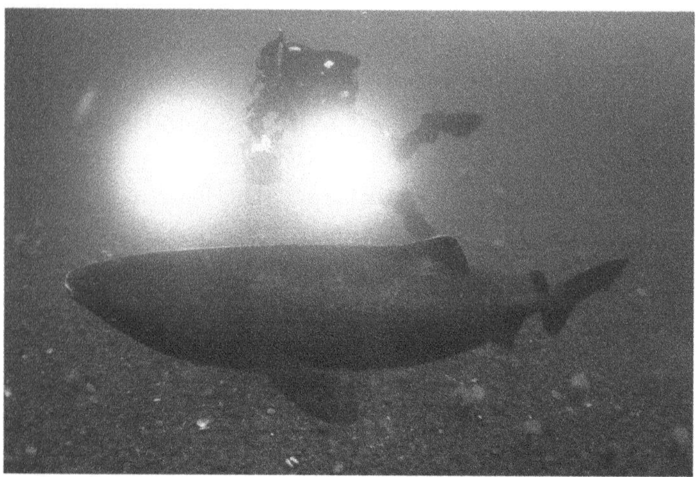

Greenland shark (Somniosus microcephalus). (Jeffrey Gallant/GEERG.ca)

The Greenland shark ranges in colour from dark brown to purplish. It has a heavy build with small dorsal and pectoral fins, big eyes behind a short, round snout and large nostrils, and a broad, thin-lipped mouth. Both upper and lower teeth are diminutive when compared with those of other large carnivorous sharks. The upper arcade consists of layers of teeth which overlap and form a dental band that is very sharp and thin, while the pointed lower teeth are separate but densely placed, with wider cusps, and probably act in a holding/gripping role during predation. The tail is moderately large. None of this suggests a dashing, top-order predator and yet the Greenland shark, as well as feeding on teleosts, squid, sharks, rays and eels, eats mammals such as seals, which suggests a turn of speed not immediately apparent in the body plan. It is very slow-growing, due to its cold water habitats.

Strange feeding habits are attributed to this shark. One is that the parasites called copepods (*Ommatokoita elongata*) which hang off the eyes of most adults—apparently causing near-blindness—act as bioluminescent prey attractors. This is unlikely to be true. Also unsubstantiated is the belief that Greenland sharks ambush caribou drinking at the mouths of Arctic rivers or attack the animals while they are crossing rivers. West Greenland longline halibut fishermen report that Greenland sharks eat their way up the longlines, so that when they themselves are finally caught their stomachs are full of hooks. While the 'gut full of hooks' claim may be an exaggeration, the shark is undoubtedly a pest to under-ice fisheries in the Arctic, as it is to the Greenland halibut fishery in Lancaster Sound in the Canadian Arctic.

A recently formed Canadian-based entity, the Greenland Shark and Elasmobranch Education and Research Group, has made many new discoveries about this intriguing predator.

According to a 2006 statement by one of the group's founders, University of British Columbia marine biologist and veterinarian Chris Harvey-Clark, 'all the papers published on the species, including magazine articles, can barely fill two shoeboxes'.5 Harvey-Clark and his colleague Jeffrey Gallant became the first divers to study and report on the behaviour of free-swimming Greenland sharks. They did so in 2003 following the discovery of a resident population in Canada's St Lawrence Estuary near the city of Baie-Comeau. (Sharks had been seen there in the past by fishers and workers at a pier construction site in the 1930s.) Their findings were numerous and revelatory. Males and females undertook vertical movements at night, regularly swimming up to the surface and back down throughout the course of the night; they exhibited classical defensive/threat postures to divers, with pectoral fins and head pointed down, the back arched, the mouth slightly open; they also exhibited curiosity about the divers. Certainly, they were not in any way lacking vision (almost none had copepods):

> Most sharks in the St. Lawrence have beautiful, crystal clear eyes and are quite visual. As you swim by, their eyes swivel and follow you, which sets them apart from the population in the arctic . . . They are likely to see relatively well [and] using a variety of other sensory modalities, they are very effective, stealthy predators and could take out an agile seal in zero visibility without alerting it . . . My take on the Greenland sharks is that they're probably like hyenas, capable of both predating and scavenging. They have lower teeth like an old-fashioned straight razor that takes a five kilogram chunk out of a whale like an ice cream scoop. But they can also suck up a large flounder

> like a vacuum cleaner. It gives you pause when you are diving with them.[6]

Such a description is far removed from the notion of a shark so slow and myopic that Inuit fishermen catch it by the tail. The St Lawrence researchers concluded that Greenland sharks were 'capable of rapid acceleration . . . were highly maneuverable and were capable of changing depth and direction rapidly'.[7] Perhaps the Greenland shark is entitled to an updating of its current scientific name. It is not to be confused with the similar-looking, equally large southern sleeper shark (*Somniosus antarcticus*) scattered in the Southern Ocean.

Bramble shark (*Echinorhinus brucus*)

This shark has a patchy worldwide distribution and is most commonly found in the eastern Atlantic Ocean, the Mediterannean Sea, southern Africa, Sri Lanka, southern Australia and New Zealand. It inhabits deep waters, to about 900 metres, towards the bottom of the continental shelf and the upper slope. The bramble shark grows to about three metres. It is a heavy-bodied shark with a pointy snout, eyes set in front of the mouth, small pectoral fins, small rearward dorsal fins, and a large sweeping tail. The very similar prickly shark (*Echinorhinus cookei*) grows to at least 2.6 metres.

The bramble shark's Latin name translates as 'hedgehog shark', with good reason. Its skin, which has a purplish tinge, is profusely covered with large, hornlike denticles. About the width of a thumbnail at the base, the denticles are tipped with cone-shaped spikes. It is thought that these spikes either generate a metallic reflection or are luminescent, for attracting prey in deep, dark waters. Bramble sharks also have a strong-smelling, thick coat of mucus over their thin skin.

Their small fins, rough, denticled skin and heavy, cylindrical shape indicate sluggish swimmers. Bramble sharks have been observed hovering almost motionless just above the ocean floor. Speculation is that it is a slow-growing and latematuring shark. Furthermore, 'almost nothing is known of the species' biology'.[8] Although the bramble shark has the bladelike teeth of a predator, it may be that its large pharynx also has a prey-inhaling role in feeding. The small dorsal fins are set well back.

The relative rarity of the species was recognised by the founding director of Australia's National Museum of Victoria, Frederick McCoy. In 1886 he took possession of a specimen that had been line-caught off Portland. He had it stuffed and mounted, labelling it the Spinous Shark. In 2004 Museum Victoria ran an 'Out of the Vaults' exhibit, in which 24 items were displayed. Visitors were invited to vote for the most popular item. McCoy's bramble shark came fifteenth (just above an FX Holden; the winner was a collection of Aboriginal stone tools called Kimberley points).

Pristiophoriformes

The sawshark, a true shark, looks like a diminutive sawfish, a batoid.

Classification

- Two genera in one family
- Approximately nine species

Biology

- Elongated snout with paired barbels
- Large spiracles behind the eyes
- Five (or six) gill slits on each side of the head
- Two spineless dorsal fins

- No anal fin
- Ovoviviparous reproduction

Habitat

- Estuarine, coastal, deep water
- No freshwater species
- Temperate and tropical waters
- Bottom dwellers over sandy substrate

Common sawshark (*Pristiophorus cirratus*)

This medium-sized shark is described as belonging to a 'minor group of harmless bottom sharks'.[9] It is endemic to the southern Australian continental shelf and upper slope. It grows to a length of about 1.5 metres, at least a third of which is the extended snout, the cartilaginous rostral saw which gives the shark its common name. There are rows of twenty or more large rostral teeth on each side of the saw, the teeth alternating between long and short. A pair of long ventral barbels, at about the midway point of the snout, once gave it the nickname the 'Fu Manchu' shark. The saw is also studded with ampullae of Lorenzini and has a lateral line. It efficiently detects buried prey, then sifts through the sand to get to the hidden meal. It also acts as a 'slice-and-dice' implement that chops up prey into bite-sized pieces. The rostral teeth are continually replaced. The sawshark has a small mouth and tiny teeth.

The rest of the shark's body is reasonably typical of a shark, being 'subcylindrical to slightly depressed'.[10] The spiracles behind the eyes are large. The gestation period is about twelve months, with about twenty live young born in a litter, the newborn measuring 30 or more centimetres in length. The pups' rostral teeth are folded back during birth to prevent injury to the mother, but then quickly become upright.

The common sawshark is fished commercially as bycatch in southern Australian waters. Together with elephant fish, sawsharks are taken by gummy shark trawlers. According to the IUCN/SSC Shark Specialist Group's 2004 Report,

> There are no useful biological data available for this species, and no assessment of the impact of commercial fishing. Although they are caught only as bycatch, the fisheries are large and have the potential to impact on the populations. Further research is needed to fully determine the status of this species . . .[11]

Sixgill sawshark (*Pliotrema warreni*)
This is one of the very few sharks to have more than five gills. Its range is restricted to the warm waters off the southern African and Madagascar coasts. A study of species interaction in the region's underwater canyons suggested that sixgill sawsharks 'are too large at birth to be easily swallowed by *Latimeria* [coelacanth], and the sharp rostral teeth of juvenile sawsharks would also discourage a hungry coelacanth'.[12]

Squatiniformes
Angel sharks superficially resemble rays because they have flattened bodies adapted to life on the ocean floor.

Classification
- One genus in one family
- Approximately eighteen species

Biology
- Dorsoventrally flattened body
- Terminal mouth
- Nasal barbels
- Eyes on top of the head

- Five pairs of ventral gill slits on the side of the head
- Winglike pectoral and pelvic fins
- Two spineless dorsal fins
- No anal fin
- Caudal fin has two large lobes
- Ovoviviparous reproduction

Habitat

- Estuarine, coastal, oceanic
- No freshwater species
- Relatively widespread in temperate and tropical waters
- Bottom dwellers in shallows down to 1300 metres or more

The angel shark—so named for its angel-like 'wings'—is one of the planet's oldest extant sharks. The lack of diversity in the order Squatiniformes is a pointer to its ancient lineage: this shark reached its evolutionary endpoint early in elasmobranch diversification. Angel shark fossils have been found dating to the Triassic Period, over 200 million years ago, not long after the demise of the xenacanths. The forepart of an angel shark is flattened and raylike, while the trunk and tail are sharklike. Angel sharks are well adapted to benthic dwelling but, despite their flatness, they are true sharks, not rays. Although their pectoral fins are greatly enlarged they are lobed; that is, they develop from the body as 'limbs', whereas the pectoral fins of rays and skates are fused as a single disc. And the angel shark's barbeled mouth is terminal, not ventral like the rays'. A few species exceed two metres in length, with most averaging about 130 centimetres.

The Australian angel shark (*Squatina australis*) inhabits

coastal waters from southern Western Australia east to New South Wales. It is a specialist ambush predator, burying itself in the sand and using the force of its lean, muscular body to lunge up at prey, which it grabs with long, thin, sharp teeth. American studies have shown that the angel shark actively travels from one location to another, where it fashions a pit in the sand and lies in wait for its prey, which includes small teleosts, cephalopods and bivalves.

Hexanchiformes

The frilled shark, placed here, is classified by some scientists in its own family, Chlamydoselachiformes. One reason is that, unlike the other hexanchids, it has a terminal mouth. (*Chlamy* means cloak or mantel, which is a description of their gills.)

Classification

- Three genera in two families
- Approximately six species

Biology

- Elongated eel-like body
- Wide terminal mouth
- Six (or seven) pairs of gill slits
- Small pectoral fins well back near the anal fin
- One small dorsal fin near the tail
- Large anal fin
- Large extended caudal fin
- Ovoviviparous reproduction

Habitat

- Cold deep waters

Frilled shark (*Chlamydoselachus anguineus*) (Plate 3)
The frilled shark grows to a length of just under two metres and is often referred to as 'primitive' because it has scarcely changed since the Jurassic shark radiation 190 million years ago. Some palaeoichthyologists believe it has an even older lineage, dating back to the Devonian well over 300 million years ago. The body of the frilled shark is elongated and muscular and it has a long, wavy tail. Emerging from a broad, reptilian head are distinctive frilly gill margins which resemble a collar or ruff. The first pair of gills are joined, that is, they form a single gill slit running under the throat.

The frilled shark is an elusive deep-water dweller, inhabiting continental shelves and slopes to depths of well over 1000 metres. This and its rarity make it difficult to study. Even so, specimens are occasionally found in shallower waters, often in the seas around Japan. The frilled shark was described and classified in 1884 by Samuel Walton Garman, the first official curator of fishes, amphibians and reptiles at the Harvard Museum of Comparative Zoology:

> For several years, [Garman] had been working on an elasmobranch that he found quite remarkable . . . This was a rare opportunity to examine the soft tissue of what appeared to be a living fossil. While many disparaging comments have been made about Garman as a systematist, no one would find fault with his anatomical skills . . . He turned his knife on the frilled shark with great success. After his initial description of the species, he published seven other papers on the anatomy of this odd creature; eventually dissecting the type specimen nearly into oblivion.[13]

The frilled shark's jaws have about 300 trident-shaped, ultra-sharp teeth set in 25 rows—ideal for impaling slippery prey such as deepsea squid. It is possible that the shark might use its elongated body to strike whiplike at prey. In January 2007, a live specimen was captured by staff of the Awashima Marine Park south of Tokyo. Although it died soon after, rare video footage was taken which clearly shows its unusual features.

Japanese studies suggest that the gestation period of the frilled shark could exceed three years. This would make it by far the longest period for an elasmobranch and, for that matter, any vertebrate.

Bluntnose sixgill shark (*Hexanchus griseus*)

Like the frilled shark, this large predator also traces its lineage back some 200 million years to the Jurassic Period. The bluntnose sixgill shark is recorded as reaching a maximum length of about five metres and weighs about 600 kilograms, although some researchers think that as a deep benthic dweller it could be considerably larger and heavier.

The shark's upper body is shaded through grey to dark brown and its underside is pale. It has a broad, muscular and heavy body, with relatively small fins, except for the elongated upper caudal lobe which propels it through the water. The single dorsal fin is as far back as most other sharks' second dorsal fin. The snout is rounded, as is the large mouth, which has six sets of large sawing teeth in the lower jaw and smaller curved single-cusped teeth in the upper jaw.

The small, green eyes have no iris, so that they can take in as much light as possible in deep, dark water. Gestation is thought to last up to two years and the resulting litters can number 100 or more. Some researchers think that such large litters may be to compensate for high mortality rates. These sharks can live for about 80 human years, but their slow, cold-water growth makes them particularly vulnerable to habitat disruption.

Despite its traditional reputation as an ancient and sluggish leviathan of the deep, the bluntnose sixgill has shown itself to be a significant apex predator with a global (but not pelagic) range perhaps exceeded only by that of the blue shark. It feeds upon all the usual prey, but also seals, dolphins and whales (blubber found in sixgill stomachs has been assumed to have been scavenged). It is reported to have an impressive burst of speed which enables it to ambush prey. Teleosts are swallowed whole, often face-first:

> I observed a 13-foot (4-metre) male Bluntnose Sixgill capture and eat a Lingcod. The shark came up along a canyon wall, swam up and over the Ling, pushing it down against the bottom with its snout. Thus pinning the Ling to the rocky substrate, the Sixgill with its tail nearly vertical spun around until it could swallow its prey head-first. This maneuver is similar to one reported in a Great Hammerhead (*Sphyrna mokarran*) feeding upon a Southern Stingray (*Dasyatis americana*) off Bimini (Strong *et al.*, 1990) and likewise seems very economical and efficient, its speed of execution belying the normally languid cruising pace of Sixgill Sharks.[14]

Bluntnose sixgill sharks are solitary, migrating vertically at night to feed. They confound perceptions again, however, by undertaking significant annual inshore migrations in order to mate and produce their live young in safe conditions. The same individuals return to the same inshore grounds year after year. Intensive studies of migratory inshore populations in Puget Sound in the northwest of the United States and Flora Islets in the Strait of Georgia, British Columbia, have provided observers with opportunities and insights into the species that would be impossible on the continental shelf and upper slope where the animals live for most of each year. Researchers at Puget Sound were able to identify the same individuals on their yearly return and noted that they appeared 'to have established movement corridors and territories that remain relatively fixed over time'.[15]

The bluntnose sixgill shark seems to be emerging from its deeps. A YouTube video shows one experimentally biting a submarine cable, before slowly swimming on. Some suggest that it may even be actively competing with the great white for prey, in waters off the Cape of Good Hope in South Africa. Unfortunately, the bluntnose sixgill shark has become targeted as the major sport fish to be had off Ascension Island in the southern Atlantic Ocean, an area that is—or was—one of its major pupping grounds.

Broadnose sevengill shark (*Notorynchus cepedianus*)

There are just two genera of sevengill sharks, commonly known as the sharpnose shark (*Heptranchias perlo*) and the broadnose shark. Together they are called cow sharks and are among the most ancient extant elasmobranchs, their body plans having undergone very little change over millions of years.

The broadnose sevengill shark has a worldwide distribution in specific, generally inshore, locations, including the western coastline of the US north to British Columbia, western and eastern South American populations, southern Africa, and from Japan south to Australia and New Zealand. It is often referred to as the Tasmanian tiger shark, because the specimen first described by the French naturalist Francois Peron in 1807 was taken at Adventure Bay, Bruny Island, which is separated from mainland Tasmania by the narrow D'Entrecasteaux Channel.

This shark grows to about three metres and can weigh in

excess of 100 kilograms. It is silvery or browny grey and is distinctively spotted. It has a liking for deep inshore channels and bays and seems to regulate its movements according to tides. In the arcuate mouth (curved like a bow) are sharp, jagged upper teeth for grasping and comb-shaped lower teeth for cutting and sawing. Its size, weight and dentition make for a shallow-water shark that is an active and powerful predator.

It is noted for its indiscriminate feeding habits, possibly equalled only by the 'other' tiger shark, *Galeocerdo cuvier.* Teleosts, sharks and rays, mammals and carrion are all eaten. Furthermore, this is one of the few shark species known to hunt cooperatively in pursuit of seals and dolphins. It is described as aggressive in its feeding behaviour. Despite this, predation on humans is virtually unknown.

THE GALEOMORPHS: MACKEREL SHARKS, GROUND SHARKS, BULLHEAD SHARKS, CARPET SHARKS

Carcharhiniformes

There is uncertainty over the precise number of genera and families in this large order, the systematics of which is described as 'the most complex and contentious of all shark groups'.[16] The Greek *Carcharhinus* means 'sharp nose' and this feature is characteristic of the requiem sharks, although many of the other families in the order are described as having short, blunt snouts.

Classification

- 48 genera in eight families
- At least 279 species

Biology

- Nictitating membrane
- Five gill slits
- Two dorsal fins
- Anal fin
- All three forms of reproduction

Habitat

- Most marine waters
- Fresh water

Pygmy ribbontail catshark (*Eridacnis radcliffei*)
This shark grows to just 24 centimetres and lives in varying depths from about 70 to 750 metres in the Indo-West Pacific, from east Africa across to Vietnam and the Philippines. It has a much greater range than other members of its genus. Dark brown in colour, it has blackish markings on the dorsal fins and prominent dark bands on the tail. It hunts along the ocean floor, preying on teleosts and crustaceans. The pygmy ribbontail catshark is ovoviviparous and gives birth to just one or two pups in a litter. This means that population numbers are difficult to sustain and the shark is exceptionally vulnerable to any kind of threat.

Swell sharks
There are at least seven known species of swell shark. They are members of the most numerous shark family, the catsharks. Rarely exceeding a metre in length, swell sharks are slow-moving bottom dwellers with large mouths containing hundreds of tiny pointed teeth, with which they prey on crustaceans and small fish.

All catsharks are small. There are several dwarf species

and the average length across the family's 160 or so species is about 80 centimetres. It's such a large family that the species' distinguishing characteristics are usually identified by their common names: flatnose catshark; brown catshark; stout catshark; longfin catshark; smallbelly catshark; Iceland catshark; Borneo catshark; longhead catshark; broadmouth catshark; fat catshark; chain catshark (which has an exquisite chain pattern on the skin); and the imaginatively named spatulasnout catshark. Most catsharks prefer deepwater slopes, although they can be found in shallower coastal ranges. The draughtboard swell shark (*Cephaloscyllium isabellum*), so named because of its skin patterning, is one of many shark species on the IUCN Red List; once abundant in waters off New Zealand, it was heavily exploited for its liver oil and, although no longer targeted, continues to suffer considerable casualties as a commercial fishery bycatch. The swell shark's common name is derived from the defence mechanism which enables it to inhale air or water when threatened, greatly inflating its stomach.

River sharks

Rare and mysterious, these are among the least known of all elasmobranchs. There are thought to be at least six species of river shark (although this remains in doubt). They are moderately stout fusiform sharks and they range in adult size from about two to three metres, with broad rounded snouts and small eyes. The speartooth shark (*Glyphis glyphis*) has been identified in a few Queensland, Northern Territory and New Guinea rivers. It is also described, however, as inhabiting inshore estuarine, brackish and low- to reduced-salinity waters. In 2003 a speartooth shark was found for the first time in a Western Australian river, although specimens had previously been taken from Queensland rivers. The speartooth is estimated to grow to two or three metres and takes its name from the fact that some of its anterior teeth have 'cutting edges confined to slightly expanded spear-like tips'.[17]

The description of the Irrawaddy River shark (*Glyphis siamensis*) is based on a single specimen found in Burma. The Ganges shark (*Glyphis gangeticus*) takes its name from the Ganges–Hooghly River system in Bengal. It is rare, but popularly considered to be plentiful because of the frequency of shark attacks in those rivers—attacks which most experts consider to be perpetrated by the unrelated bull shark.

The Australian species of river sharks for many years had the tantalisingly uncertain names *Glyphis* species A, B and C. The description of *Glyphis B* had been based on a single specimen, until in 1998 a team from the IUCN Shark Specialist Group, working with local fishermen, rediscovered the species in a few rivers and named it the Borneo river shark. More recently *Glyphis C* has been renamed the northern river shark, *Glyphis garricki,* a species restricted to some freshwater systems of northern and Western Australia and New Guinea. This followed research by the CSIRO'S Wealth from Oceans Flagship project.

Why are these sharks so rare? It would be tempting to reply that sharks are not adapted to live in freshwater systems, so river sharks have always been rare; but some rays and sawfishes are exclusively freshwater, and the bull shark, a large predator, is equally at home in the sea and as far inland as it can get. The more accurate response is depressing: 'Once prominent in tropical river systems, fishing pressure, pollution and habitat destruction through overdevelopment have greatly reduced the natural populations of freshwater sharks and rays'.[18] These sharks' slow growth rates, low reproductive levels and adaptation to an exclusive environment over a great period of time mean that even minor changes to that environment can fatally disrupt a population.

It may be that river sharks are adapted to fresh water with particularly low oxygen levels, a deterrent to bull sharks—which would prey upon both juveniles and adults—entering such rivers.

Reef sharks (Plate 4)

In stark contrast to river sharks, the four species of reef shark have been well studied. Coral reefs and atolls, in sunlit nutrient-rich waters, support a great variety of living organisms and there is no shortage of prey for local predators. So why are there so few shark species? The fact is that reef sharks are numerically abundant, but each species has evolved to occupy a specific ecological niche, thus minimising intraspecies competition.

The whitetip reef shark (*Triaenodon obesus*) grows to about 1.7 metres in length and weighs about 45 kilograms. It is widespread in the central and western Pacific Ocean, with populations also in Asia and Africa. It is lean, with a flattish head and a broad snout, and prominent white tips on its dorsal and upper caudal fins. It has small, smooth-edged, thin, sharp teeth in the bottom jaw. The whitetip reef shark is generally described as 'sluggish' and during the day can be found resting in sandy caves or under coral overhangs. Unlike the pursuit predator, the whitetip reef shark doesn't rely on speed. It uses its flattish head to search out the bony fishes, crabs and octopuses which hide in coral reef cracks and holes. The shark thrusts its head into its prey's hiding place, driving itself inward by twisting and turning its body. It dislodges rocks, breaks off pieces of coral (frequently cutting itself in the process) and sometimes squirms right through the gap where the prey had been.

A common, gregarious species which lives in clear, shallow waters and is not aggressive towards humans, the whitetip reef shark is one of the most studied of sharks. One American study of nine whitetip reef sharks at a Cocos Islands reef managed to capture their mating behaviour on film, and the resulting footage became the first to show males who were unsuccessful with the female of their choice ejaculating into free water and swimming off with mouth agape. This study also offers the first detailed visible evidence of the size and precise functioning of the male siphon sacs.

Reproductive behaviour of these sharks is unusual. Multiple males attempt to copulate with one female, who uses a range of avoidance techniques, such as tucking in a pelvic fin to avoid it being grabbed by a male, and bending her body to keep her cloaca away from the claspers—strategies thought to indicate deliberate mate selection on the part of the female. Group courtship is common in some other species of sharks but had not been seen in whitetip reef sharks. Once a female has accepted a male, the act itself appears awkward, but also graceful:

> The two sharks tumbled in copula down an underwater cliff face, with the male showing rapid caudal maneuvering that brought the pair to rest in a vertical head-down orientation on a rocky wall. Once they came to this position, the male ceased his swimming movements and began rhythmic but variable thrusting of his pelvic region towards the female. They remained there for several seconds before falling and coming to rest twice more, each time maintaining parallel, head-down positioning . . . At least four peripheral males

> with mature claspers circled the mating pair throughout the event. They nudged the female's free pectoral fin with their rostrums several times, but did not grasp it . . .[19]

The blacktip reef shark (*Carcharhinus melanopterus*) is also widespread in the tropical waters of the Indo-Pacific, its common name derived from the distinctive black tips of the first dorsal fin and lower caudal lobe. It is similar in size to the whitetip reef shark, reaching a length of about 1.8 metres and weighing up to 45 kilograms. It is a streamlined, agile shark with classic countershading: grey upper body and white lower body, with a dark stripe along each flank. The blacktip is a pursuit predator and often hunts in a pack, rounding up shoals of fish which are then consumed in a feeding frenzy. Like the whitetip reef shark, the blacktip has a very small range, and frequently inhabits shallow mangrove and lagoon waters, its dorsal tips exposed as it cruises. Along with its diet of fishes, crustaceans and cephalopods, a northern Australian study found that almost a quarter of blacktip reef sharks' stomachs contained terrestrial snakes—clear evidence of its extremely shallow water lifestyle.

Unfortunately, any healthy juvenile blacktip unlucky enough to be netted may be kept alive to gratify the human desire to possess the unusual and rare:

> Amongst the 'real shark' looking sharks offered in the aquarium trade (other than those Nurse Sharks, Leopards, Epaulettes, Bamboo, Catsharks . . . that spend so much time 'just sitting on the bottom') is the Requiem Shark (family Carcharhinidae) member *Carcharhinus melanopterus* . . . Unfortunately, this shark is entirely unsuitable for home aquarium use, requiring a pool-sized enclosure (thousands of gallons)

> to do well. Though folks can/do keep small specimens of the Blacktip Reef Shark, aka the Reef Blacktip Shark in much smaller systems for a time, invariably these are short term successes, with the specimens almost always dying 'mysteriously', crashing into the tank's side, or jumping out. As with all the other 49 species of Requiem Sharks, you're encouraged to go visit them in the ocean or Public Aquariums.[20]

The grey reef shark (*Carcharhinus amblyrhynchos*) shares the same Indo-Pacific habitat as the whitetip and blacktip reef sharks, and is similar to the latter except that its upper body is silvery-bronze. The teeth are triangular and serrated in the upper jaw and narrow and generally smooth-edged in the lower jaw. Social and tending to school, it is not only the largest of the three, exceeding two metres in length, but probably the fastest, with a top speed approaching 50 kilometres per hour. Its range takes in coral shallows as well as deeper areas near the drop-offs to deep water. Shark researcher Leonard Compagno has studied the relationship between the three species:

> . . . [the grey reef shark] shows microhabitat separation from the blacktip reef shark; around islands where both species occur, the blacktip occupies shallow flats while the grey reef shark is usually found in deeper areas, but where the blacktip is absent the grey reef shark is commonly found on the flats . . . [The grey reef shark] complements the whitetip shark, as it is far more adept at catching off-bottom fish than the whitetip, but the latter is far more competent in extracting prey from crevices and holes in reefs.[21]

Compagno also describes the grey reef shark's threat display, as investigated by behavioural researcher Donald R. Nelson:

> This consists of an exaggerated swimming pattern in which the shark wags its head and tail in broad sweeps, arches its back, lifts its head, depresses its pectoral fins and sometimes swims in a horizontal spiral. The display varies in intensity from merely a component of flight from the accosting diver to a series of figure-8 loops in front of the aggressor. Using a small shark-shaped 'Shark Observation Submersible' to approach grey reef sharks, Dr Donald R. Nelson was able to elicit threat display from the sharks while other divers filmed the behaviour from a safe distance. When persistently approached by the sub, some of the displaying sharks fled, but a few terminated the display and attacked the sub at high speed, biting one or more times and then fleeing. The speed of the attacks and the damage to the sub was impressive, and is a mute warning that these sharks should be treated with respect and not cornered or harassed by divers. The threat-display behaviour of this shark is thought by some researchers to possibly intimidate potential predators on it.[22]

The Caribbean reef shark (*Carcharhinus perezi*) can grow to a little under three metres and shares many of the physical and behavioural characteristics of other reef shark species. It is, however, distinguished by its restricted geographical range—the reefs and adjacent waters of eastern South America north to the east coast of Florida, the Gulf of Mexico and the Bahamas.

Caribbean reef sharks played a central role in a shark-tracking project which used surgically implanted acoustic transmitters to monitor the movements of shark populations and to determine the threats posed to the animals by commercial fishing. This Pew Institute for Ocean Science project took place over four years at Glover's Reef Marine Reserve (GRMR), an atoll 45 kilometres off the coast of Belize, home to twelve species of sharks and rays.

A Marine Protected Area (MPA) is one in which fishing is prohibited. Obviously, sharks have no concept of such boundaries and an MPA is of little use if it doesn't protect the species most at risk, hence the importance of tracking. The Pew project also set out to determine to what extent the coral itself is dependent for its health on the animals, including predatory sharks, which share its habitat.

Results showed a strong tendency to 'site fidelity', but also confirmed the unpredictability of shark behaviour. Caribbean reef shark 3348, an adult male, was tracked as follows:

> Although he routinely patrols the eastern reef slope and visits the east side of Middle Caye on an almost daily basis, 3348 occasionally disappears from Glover's Reef. On one such trip, for about four days in July 2004, he swam across deep, open water to Lighthouse Reef (30 km to the northeast of Glover's Reef) . . . Despite this, 3348, like all other Caribbean reefs tracked to date, spends many days at GRMR and does not seasonally emigrate from this location for any length of time.[23]

Their unpredictability is one of the characteristics which make sharks so fascinating to researchers, but it also complicates the already difficult task of promoting, let alone enforcing, conservation in the oceans.

The false catshark (*Pseudotriakis microdon*)

This is a rare ground shark which grows to about three metres. Although sometimes found in moderate depths, it generally lives in very deep waters, to about 1500 metres. Its usual range is in the northern hemisphere waters, although it is occasionally found in the Indian and Pacific Oceans. The false cat is named for its elongated cat-like eyes. It is dark pinky brown in colour, with a long snout, large spiracles and hundreds of rows of tiny teeth in a large mouth with a huge gape. The low first dorsal fin extends for a long way along the back, but the fins, musculature and skin are all flabby, indicating a sluggish swimmer. The pectoral fins are small and lobe-like. The stomach contents of specimens taken in the Pacific included deepwater eels, grenadiers, mackerel, lanternfish, squid and octopus, but also pufferfishes, which are normally restricted to shallow waters. Researchers speculate that the sharks may have scavenged dead pufferfishes that had sunk to the ocean floor.

In 2004 a false catshark was recorded for the first time in eastern Australian waters, by a commercial fishing vessel undertaking exploratory species targeting in the Coral Sea. It was a mature male, measuring 277 centimetres. Interestingly, in that same year a slightly larger female became the first recorded in nearby Indonesian waters when it was taken near the island of Lombok.[24]

The lemon sharks

The lemon shark has been intensively researched, because it is easily kept in captivity and not difficult to study in the wild. The Bimini islands off Florida's coast are home to the Bimini Biological Field Station, whose biologists have for many years specialised in lemon shark research, because the species spends its entire life cycle of about 25 years in the local waters. This whaler shark is fusiform and grows to about three metres, weighing 120 kilograms or more, and is in many respects a 'typical' shark. It has prominent, firm, broad-based pointed fins and a large powerful tail. The body is variously described as moderately stout, stocky, or robust, all of which apply to the whaler clan. (This uniform similarity is why it is difficult to differentiate between many of the whaler sharks.) Lemon sharks have a widespread global distribution, fringing the continental shallows of the western and central Pacific Ocean, the Indian Ocean, the Atlantic and the Gulf of Mexico. There are currently two recognised species, *Negaprion brevirostris,* common in the western Atlantic, and *Negaprion acutidens,* found in Australasian waters. These sharks have a yellowy-brown dorsal colouring with a paler yellow underside. Juvenile lemon sharks are born in litters of up to about fourteen, the pups then developing in mangrove shallows, which are both rich in prey and relatively safe from predators. The eyes of juvenile lemon sharks are adapted to cope with the tannin-coloured water of these shallows, which alternates between bright and dark. And research on lemon shark eyes has had broader results:

> Perhaps the most ecologically significant discovery about shark retinas was revealed in a fascinating 1991 paper by Robert Hueter. Hueter discovered that the Lemon Shark has a broad, horizontal band that lies across the equator of the retina and is disproportionally rich in cones. Based on a similar retinal band in the Lion (*Panthera leo*), Cheetah (*Acinonyx jubatus*), Thompson's Gazelle (*Gazella rufifrons*), Wildebeast [sic] (*Connochaetes taurinus*), and other mammals of the African plains, this so-called 'visual streak' probably grants the Lemon and other sharks a particularly clear view of the underwater horizon. As a shark's potential prey, rivals, and mates are most likely to first appear on the hori-

> zon at the limit of visibility, the adaptive (survival) value of the visual streak is easy to imagine.[25]

Unfortunately, worldwide destruction of mangrove areas, primarily for coastal development or aquaculture farming, is affecting lemon sharks, as well as every other marine species reliant upon these critical bodies of water.

Juveniles swim together and this group behaviour persists into adulthood. As the animals mature they move into the open waters, with a home range of some hundreds of square kilometres. The lemon shark is a tropical water apex predator, fast in pursuit with long, thin, sharp, finely serrated teeth ideal for grabbing a variety of prey: teleosts, rays, smaller sharks and seabirds resting on the ocean surface.

Oceanic whitetip shark (*Carcharhinus longimanus*)

This pelagic species averages about three metres. The heaviest recorded specimen to date weighed 167.4 kilograms. It is one of the most abundant, cosmopolitan and widely distributed pelagic sharks. Highly migratory, it is found in circumglobal warm waters from as far south as south-western Australia to northern Atlantic waters off southern Canada. The tips of the huge dorsal, pectoral and upper caudal fins are generously splotched with white (hence its common name), contrasting with the evenly denticled bronze-grey skin of the upper body. *Longimanus* translates as 'long hand', a good description of its winglike pectoral fins.

The oceanic whitetip has a short, broad snout. Its powerful jaws contain about 60 teeth. Those in the upper jaw are broad at the base, heavily triangulated and serrated, those in the lower jaw thinner and sharper, with little or no serration. The shark seizes large teleosts such as tuna, barracuda and marlin by thrusting the gaping jaws forward, impaling the prey on the lower teeth while the upper teeth engage in sawing and tearing actions to remove a large chunk of flesh. The oceanic whitetip will swallow smaller prey whole and, although solitary, will occasionally hunt as a pack, herding school fish into tight balls and feeding communally. They have also been observed associating with shortfin pilot whale pods which are expert at hunting one of the whitetips' prey, squid. They regularly follow tuna fishing boats, which likewise lead them to prey. They also have a predilection for garbage dumped from ships.

The most common adjectives used to describe the whitetip are slow, lazy, aggressive, dominant, inquisitive and persistent. Certainly, with its bulky, slightly humpbacked body and large, broad fins, the whitetip is less slimline than the truly fast big sharks: 'Reports have described swimming behavior in open waters at or near the surface of the water as moving slowly with the huge pectoral fins spread widely'.26 This is not to say, however, that it is a 'slow' shark. Like an eagle or vulture gliding in a thermal, the whitetip conserves its energy, and has plenty of muscle and tailpower in reserve to chase and strike.

The common and probably mistaken belief is that the great white shark is the number one maneater. The oceanic whitetip, whose domain is the upper pelagic layer, is an opportunistic predator and invariably the first to arrive, and in greatest numbers, at shipwreck sites. It is highly likely that over the centuries whitetips have eaten many more shipwreck and other marine disaster victims than any other species of shark.

Galapagos shark (*Carcharhinus galapagensis*)

The shark's common name suggests a particular affinity with the Galápagos group of islands in the Pacific Ocean west of

Chile, famed for their uniquely isolated flora and fauna and the double geographic rarity of being sandless desert islands. While this requiem shark is common in that area, the name derives only from the fact that a specimen caught there early in the twentieth century was the first of the species to be described. It grows to about three metres and weighs about 85 kilograms.

Like the oceanic whitetip, the Galapagos shark is circumglobal in tropical waters but the similarities end there. The Galapagos shark is slimmer, with straighter, more pointed dorsal and pectoral fins, a longish snout and uniformly brown-grey skin with white underside. Its worldwide distribution is described as 'patchy'. This is because it has evolved as an oceanic island specialist, preferring the clear, shallow inshore waters often characterised by rocky bottoms and fast-flowing currents found around deepwater islands. It generally feeds on bottom dwellers—teleosts, cephalopods, sharks and rays—and has similar defence/threat postures to those of the grey reef shark. Those sharks which inhabit the waters around the Galápagos group also prey on sealions and marine iguanas.

Although these sharks have been tracked swimming between islands, it seems that their home ranges are based firmly on islands, some well known—Hawaii, Lord Howe, Madagascar, Cape Verde, Bermuda and the Virgin Islands—others less so, such as Kermadec Island off New Zealand, the Revillagigedo Islands off Mexico, Sao Tome and Principe off Nigeria, the Tuamoto Archipelago, Clipperton . . . not to mention St Peter and St Paul's Rocks.

The Mid-Atlantic Ridge, the planet's longest mountain range, occasionally breaks the ocean's surface to form islands such as Ascension, St Helena and the outcrop of St Peter and St Paul's Rocks. Brazilian national territory, the Rocks consist of nine tiny outcrops totalling less than two hectares. They are the peak tips of a 4000-metre submarine mountain equidistant between the east coast of South America and the bulge of western Africa. Some ten million years old, they are rich in magnesium and iron, a visible manifestation of Earth's active mantle. Their geological volatility and the wave action around them mean that they are changing shape constantly, and the life forms associated with them are themselves in a permanent state of adaptation. Algaes grow on corals and pridotite (a non-volcanic rock) that form near-vertical undersea dropoffs. The outcrops support only mosses, primitive grasses, insects and spiders, and were described by Charles Darwin in 1831 as 'from a distance of a brilliantly white colour . . . owing to the dung of a vast multitude of seafowl'.[27]

Researchers are not sure why the Galapagos shark favoured these rocks. An early twentieth-century expedition by the Scottish National Antarctic Expedition, led by William Spiers Bruce, recorded significant shark activity: 'Bruce and his party had hoped to land but the swell was too strong. [Medical officer] Pirie jumped ashore but fell into the sea. He managed to scramble back aboard as the crew fended off sharks with their oars'.[28]

In 1980, a Cambridge Expedition survey found that the Rocks 'support one of the densest shark populations in the Atlantic Ocean . . . Since [American navigator Amano] Delano's visit in 1799 . . . visitors to St Paul's have remarked on the extraordinary number of sharks surrounding the Rocks'.[29] The same researchers recorded that Galapagos sharks were 'unusually common',[30] making them the dominant resident predator. But just twenty years later, a Brazilian survey found a dramatically altered ecosystem. During the course of four expeditions between 1999 and 2001, involving 47 days of diving to 62 metres, and rock pool surveying, researchers observed many fish species noted in the Cambridge survey, but they did not see a single Galapagos shark.

Why would the apex predator at one of the planet's most remote landforms vanish from its pre-eminent niche? The Brazilian surveyors' gloomy assessment was that while the past abundance of Galapagos sharks could be 'partly attributed to the lack of fishing activity',[31] since the end of the 1970s 'fishing pressure has increased greatly, and sharks are now targeted due to the high commercial value of their fins. The pelagic fishes on which sharks feed are also targeted by the fishing industry and this may also have contributed to an apparent population decline of *C. galapagensis*'.[32]

Blue shark (*Prionace glauca*)

The blue shark was one of only two shark species recorded in the Brazilian survey of St Peter and St Paul's Rocks (the other being the silky shark (*Carcharhinus falciformis*)). Yet far from being rock and reef specialists these two sharks, along with the oceanic whitetip, make up the trio of tropical and temperate pelagic requiem sharks considered to be the most abundant big sharks of their kind.

The blue shark's circumglobal range is vast, extending considerably further south and north than that of the oceanic whitetip. This shark is also considered to be the greatest traveller, undertaking migrations of hundreds and even thousands of kilometres. Adults grow to almost four metres, and weigh up to 200 kilograms. The website of the Florida Museum of Natural History claims that 'they are rumored to get as large as 20 feet [six metres]'.[33]

The blue shark is brilliantly coloured, being shiny indigo

blue on its upper surface, paling to an even white underneath the slender body. The snout and head are streamlined, the eyes large and white-rimmed, the dorsal fin modest, the upper lobe of the tail fin elongated and slender, and the pectoral fins long and winglike. This graceful and elegant shark is described as an 'open-ocean glider'.[34]

The blue shark eats a broad range of prey, mainly relatively small teleosts and cephalopods, although as an open-ocean opportunist it will take shipwreck victims. One adaptation to existence in an open-ocean environment where food can be scarce is the presence of gill rakers that prevent even tiny food items escaping. It is not uncommon in coastal waters, making it unusual as a species that is both a fringe littoral and oceanic–epipelagic inhabitant.

The blue shark's inshore feeding strategies follow prey movements. The Hudson River, flowing through New York City, continues on as the submerged Hudson Canyon, a kilometre-deep rift in the continental shelf off the coast. Between 1979 and 1996, a seminal project tagged feeding blue sharks active in this inshore water. During the day the sharks swam slowly near the surface, but periodically dived to depths approaching 400 metres, before surfacing again. Sharks were observed to make up to six dives during daylight hours, with a diving technique formally described as 'remarkably regular vertical oscillations . . . with some [sharks] dropping very rapidly'.[35] The shark's return to the surface may have something to do with thermoregulation, given the temperature differences between the water layers.

Blue shark migrations are associated with their reproductive cycle. Tagging studies revealed populations mating in waters off the Caribbean Sea and east coast of the United States. The pregnant female stores the male sperm in sacs in her reproductive tract and sets off towards Europe, planing through the ocean in the currents of the Gulf Stream and the North Atlantic Gyre—the aquatic equivalent of albatrosses circumnavigating Antarctica in the winds of the Roaring Forties. At some point in the journey, possibly as long as twenty months after the act of copulation, the female's eggs are fertilised. The timing is perhaps linked with the female's own maturation process. Litters that can exceed 100 live young are born in pupping waters off the west coast of Europe. The female then continues on to the north-west coast of Africa, in order to link up with currents that will return her to the Caribbean Sea. Such an incredible journey could total up to 15 000 kilometres, and may take the shark about two years to complete.

Blue shark pupping grounds are also known in the South-west Atlantic, the Northern Pacific and the South-west Pacific, and there may well be others. Why such a diverse range? There are multiple reasons, according to longtime blue shark researcher John Stevens: 'The oceanic environment is huge and food resources are often scarce in nutrient-poor open ocean waters. Pupping areas are often in more productive zones so that there is sufficient food to grow the pups rapidly out of the predation window, and separating different components of the population by area also reduces intra-specific competition'.[36]

Bull shark (*Carcharhinus leucas*) (Plate 5)

This is the last of the requiem sharks to be described here. The bull shark is the only large saltwater predator also capable of living in freshwater rivers and lakes. It grows to about four metres and weighs in excess of 200 kilograms—significantly heavier than the longer, leaner oceanic whitetip. The bull shark is variously described as stout, robust, unpredictable and pugnacious: a mixture of physiological and behavioural characteristics that underscore its reputation as a vigorous warm-water predator and scavenger.

The bull shark has a noticeably short, blunt snout (its French name is *requin bouledogue*) and large, triangular serrated teeth. The eyes are quite small, well-adapted to shallow, often murky or silty water in which eyesight is less important than other senses in detecting food. The heavy, fusiform body is propelled by a large, powerful tail, the first dorsal fin has a large base and a distinct triangular form, and the pelvic fins and second dorsal fin are also well-developed. The fact that the bull shark inhabits marine, estuarine and freshwater systems means that it takes a range of prey, with a concentration on teleosts, rays and smaller sharks (including its own kind).

The bull shark has a notorious reputation as a maneater, partly because, as an inshore and freshwater species, it comes into contact with people more often than other large predators, and also because it is a somewhat indiscriminate forager. Many shark researchers believe that attacks automatically attributed to great whites are just as likely to be the work of the bull shark. The dentition of the two species is similar and analysis of the bite marks can be inconclusive.

Like most shark species, there is much still to learn about the bull shark. Richard Pillans of the University of Queensland, in his study of the physiological ecology of bull sharks in the Brisbane River, determined that the animal's liver, kidneys, gills and rectal glands modify the body's salt levels according to the degree of salinity in the water around it. The kidneys maintains equilibrium by retaining or discharging salt-concentrated urea and the gills undergo a cellular alteration to accommodate the changing salinity of the water flow.[37]

Pillans also identified bull shark nursery grounds in Queensland estuarine systems; the first known in the South Pacific.

Other studies have concentrated on why these sharks move between different habitats. Bull sharks resident at the Bahamas for nine months of each year mysteriously vanish for the remaining three months. A pilot conservation program, Project Aware, tracked a number of individuals. One adult female headed directly to Florida and entered a freshwater system that bull sharks use as a nursery ground. Researchers are also studying a Pacific Ocean population of bull sharks, using data about their movements to understand more about their behaviour. The bull sharks are not always as fearsome as their reputation:

> The bull sharks we study off Viti Levu in the country of Fiji stay deep at 30 to 40 meters depth in the water column. We also wanted to target specific individual sharks which could yield the most useful data. Since we did not want to traumatize the sharks through conventional scientific catching techniques to deploy our tags we chose to intimately enter the bull sharks' world and interact with them face to face. Swimming among two dozen or more large bull sharks at 30 meters is a life changing experience and we were immediately awestruck by their beauty, majesty and grace. They would allow us to approach them and attach the tag into their dorsal musculature just below the first dorsal fin with a custom designed tagging stick. Over the course of the year 2004 we equipped a total of 11 personally selected adult bull sharks with satellite tags . . .[38]

The broad and fast-flowing rivers of the south-east coast of Africa carry great amounts of silt; their rivermouths are often muddy-brown, turbulent and home to Zambezi sharks—as the bull shark is known locally. The Zambezi shark is considered to be a major killer off the popular swimming beaches of the South African province of Natal. This may be so but, according to a local dive operator:

> Diving with Bull Sharks in South Africa and Mozambique is quite safe—if you respect the sharks and stick to the rules. No scuba diver has ever been attacked by a Bull Shark while diving, and we intend to keep it that way. We expect to start seeing Bull Sharks from September, with the most sightings from November to May. Sightings in other months are rare. The animals are not fed, baited or conditioned at all—the Bull Sharks are completely wild. The Bull Sharks are curious of nature—therefore we often see them as soon as we hit the water because they have been waiting to find out if it is a fishing boat or not (clever buggers). The splash of divers doing backward rolls into the water causes the Bull Sharks to scatter at first. They often come very close to inspect the divers on both the descent and ascent, but easily lose interest.[39]

Bull sharks are found in many freshwater systems. They travel enormous distances inland in rivers such as the Mississippi and Amazon. A Caribbean population that mates and breeds near the mouth of Nicaragua's San Juan River regularly migrates 200 kilometres upriver, negotiating three sets of large rapids, into Lake Nicaragua. An individual takes a week or more to make the journey, the purpose of which is thought to be associated with the breeding cycle, because the sharks do not form a permanent population in the lake (unlike the entirely freshwater Nicaraguan sawfish). Lake Nicaragua averages a shallow thirteen metres, which suits the bull shark.

The bull shark is generally thought of as 'slow', perhaps because of its bulky body, but it has an impressive turn of speed. The cartilage of its jaws is unusually rigid (due to a concentrated calcification process) and in attack, its mouth is almost squarely open. This makes it extremely efficient at tackling and extracting great chunks from large prey. It's a feeding strategy supported by an unusually muscular neck/nape, a musculature which increases with age. These physical characteristics have led the bull shark to be compared to terrestrial predator–scavengers such as the Tasmanian devil, famous for its ferocious-looking gape and large, bullish neck and head on a proportionately small body.

Hammerhead sharks (Plate 6)

There are nine species of this shark, some of which are quite small, but the largest, the great hammerhead (*Sphyrna mokarran*), grows to about six metres and weighs 450 kilograms or more. Hammerheads are widespread in tropical and temperate waters, from coastal shallows to semi-oceanic depths of about 300 metres. During the summer they migrate, moving polewards into cooler waters. They are not oceanic.

It is both apposite and opposite to link hammerheads and threshers, the exceptional tails of the latter matched by the heads of the former, which supply both their scientific name (*sphyrna* means 'hammer') and common names: the winghead shark, the bonnethead shark, the mallethead shark, the scoophead shark and the scalloped hammerhead.

S. mokarran has a powerful fusiform body, a tall, pointed dorsal fin, big pelvic fins and a large upper caudal lobe. The dermal denticles are closely spaced, and the mouth a semicircle set well back underneath the flattened head. The teeth are relatively small, triangular and serrated. It is an active predator, with a wide prey base and a preference for stingrays. Litters of live young may number more than 40, but great hammerheads are solitary animals, unlike other hammerhead species, such as the scalloped hammerhead, which school in large numbers: groups numbering in the low hundreds are not uncommon.

In a species family known for its unusual heads, *S. mokarran* is so unusual that its head has acquired its own generic name, a 'cephalofoil', defined as the lateral extensions of the head of a hammerhead. Descriptions of *S. mokarran* refer to 'its nearly rectangular head';[40] its 'T-shaped head';[41] or the fact that the 'front margin of the head is nearly straight . . . [with] eyes located at the tips of laterally expanded blades'.[42]

These sharks' widespread distribution and preference for shallow waters mean that researchers have been able to work with them over many years, but there remains a degree of uncertainty over the precise evolutionary solution achieved by the head shape. University of Hawaii researchers, using electrically charged dipoles to simulate prey, hypothesised that the 'unique head morphology of sphyrnid sharks might have evolved to enhance electrosensory capabilities',[43] but concluded that 'although the sphyrnid head morphology does not appear to confer a greater sensitivity to prey-simulating dipole electric fields, it does provide (1) a greater lateral search area, which may increase the probability of prey encounter, and (2) enhanced maneuverability, which may aid in prey capture'.[44]

One or more of four possibilities—binocular vision, stereo smell, electroreception, manoeuvrability—remain the accepted explanations for the shape. The latter is intriguing, posing to the University of Hawaii researchers the question, 'does the hammerhead help this shark to turn on a pinhead?'[45] Such precision could do for the hammerhead what the tail may do for the thresher shark. Furthermore, the enormous head of the winghead shark (*Eusphyra blochii*) is about half the length of this shark—the same proportion as the thresher's tail to its body. The common name of this species scarcely requires explanation: 'In practice, the head is transformed into a wing able to support and aid changes of direction.'[46] This shark grows to about 1.8 metres.

Schooling behaviour also remains something of a mystery. Hammerhead schools can number in hundreds and form part of distinct behaviour patterns. They gather during the day, generally at seamounts. Their interactions are highly social, with numerous displays, some of which include use of the hammer as a butting instrument. There is a dominance hierarchy, by which the largest females occupy the middle of the school and are often approached by males who otherwise seem not to be part of the group. It would appear, therefore, that mate selection is one function of the school. Towards evening, the school disperses and each animal spends a solitary night feeding in deeper waters. They then return to the same seamount the next day. The schools move from one seamount to another, although they are not unique fish in this. According to the longtime hammerhead shark researcher Peter Klimley, whose work has focused on the scalloped hammerhead, 'There is a whole assemblage of species that move north and south via seamount stepping stones'.[47] As a highly visible marine species whose numbers are under threat, the migratory habits of schooling hammerheads are being closely monitored by Klimley and his team at the University of California Davis. The team is using electromagnetic fields associated with seamounts to map their journeys—that is, creating geomagnetic shark roads—an approach Klimley hopes will prove the critical importance of seamounts and encourage governments to turn the areas around them into marine reserves.

Tiger shark (*Galeocerdo cuvier*)
This requiem shark is the only member of its genus, in contrast for example to the closely related *Carcharhinus* genus which contains about 30 species. There is some doubt about its maximum length, but specimens measuring 7.5 metres and weighing over 800 kilograms have been recorded.

The tiger shark is widespread in tropical waters and is the primary apex predator in the planet's warm coastal waters. Newborn and juvenile tiger sharks have distinctive black vertical bars and spots, which become less pronounced in adulthood, with the skin a uniform grey. The underside is typically white. The adult shark's snout is usually described as blunt, wedge-shaped and robust; a shape that allows the animal to turn its head rapidly and shake it vigorously from side to side when feeding. The eyes are large and the mouth huge, almost as wide as the head. The teeth in both jaws are broadbased, flat, heavily serrated and asymmetrical, or 'cockscomb'. The cartilage of the jaws is strengthened with multiple layers of calcium salts called tesserae, normally found in a single layer in shark jaws, but the force of the bite of the tiger shark—and the great white—requires this strengthening.

As the shark matures, its body forward of the dorsal fin continues to bulk up. The first dorsal fin is large and upright, and acts as a fulcrum, helping the shark to pivot and turn quickly; the long, downswept pectoral fins act as wings in providing lift; and the elongated tail generates powerful forward thrust.

The tiger shark is found in a variety of shallow water habitats such as harbours, estuaries, lagoons, river mouths and inshore reefs, where it feeds at night on surface, midwater and bottom prey. It has also been recorded at depths of at least 150 metres, which means that it has a wide prey base. It is a migratory species, moving into cooler waters in summer. The tiger shark is the only ovoviviparous requiem shark and the females give birth in shallow pupping grounds.

The tiger shark has been condemned variously as the 'dustbin' of the sea because it eats 'everything', and as a ferocious maneater lurking in shallow murky waters. Neither charge is true. Like any other shark, it has evolved slowly over a great period of time to take full advantage of its particular niche in the marine ecosystem. Certainly, its diet is cosmopolitan:

> The stomach contents of 811 specimens of *G. cuvier* caught in the Townsville area [in shark nets] were examined. Of these, 31.2% were empty . . . The main prey groups were teleosts and sea snakes, but crabs, turtles and birds were also important. Other prey groups included dolphins, dugongs, sharks, rays and squids. A number of items of garbage were also recorded, including plastic bags, aluminium foil, clothing, a sugar bag, prawn netting and kitchen scraps . . . The proportion of teleosts in the diet decreased with increasing shark size. Other food groups that decreased with increasing shark size were squids and birds. Concomitant with the decrease in these groups was an increase in the occurrence of turtles, crabs and sharks in larger individuals.[48]

This is not the diet of a true scavenger; rather, it is euryphagic (eating a wide range of foods). The oceans are cluttered with human refuse, any amount of which may have been ingested by the prey rather than the predator. This does not mean that the tiger shark doesn't eat garbage: behaviourally it is described as curious and assertive; and feeding is a major component of behaviour. The significant fact in the quote is that as tiger sharks grow they eat a greater proportion of turtles. This relates to their complex dentition. It is increasingly believed that tiger sharks' teeth have evolved to take advantage of large turtles (adult green turtles reach 100 centimetres and weigh about 150 kilograms). The tiger shark's heavily serrated teeth 'are reminiscent of a can opener'.[49] When the tiger shark clamps the turtle between its teeth and shakes its head from side to side, the teeth saw through the carapace top and bottom.

Concurrent but unrelated Australian studies are adding to the meagre understanding of this unique shark. A tagging program at Raine Island off Cape York Peninsula, north-eastern Australia, has revealed a new understanding about the movement patterns of the species. The tagging site was chosen because tiger sharks congregate in large numbers around the island in summer, in order to feed off its turtles and seabirds. Researchers attract the shark alongside their boat with bait, then clamp its tail with a 'shark claw' attached to a rope and a float, the drag of which, in a modern echo of the shark caller's technique, quickly immobilises the animal without in any way harming it. The tag (a satellite transmitter, long-life batteries, aerial and saltwater switch all encased in a waterproof resin pod) is then attached to its dorsal fin, and the signals it transmits enable the shark's movements to be tracked.

On the western seaboard of the Australian continent, there is extensive research into the tiger shark in Shark Bay (where Dampier described his specimen 300 years ago). The bay's seagrass beds are major feeding grounds for dugongs, green turtles, dolphins, cormorants and sea snakes, all prey items of the tiger sharks that enter the bay. The program is associated with the Shark Bay Ecosystem Research Project, led by the Heithaus Lab of Florida International University.[50] The tiger shark research had a specific aim:

> Sharks are top predators in many marine communities, yet no studies have quantified the habitat use of large predatory sharks or determined the factors that might influence shark spatial distributions. We used acoustic telemetry and animal-borne video cameras (Crittercam) to test the hypothesis that tiger shark (Galeocerdo cuvier) habitat use is determined by the availability of their prey. We also used Crittercam to conduct the first investigation of foraging behavior of tiger sharks . . . Although there was individual variation in habitat use, tiger sharks preferred shallow seagrass habitats, where their prey is most abundant. Despite multiple encounters with potential prey, sharks rarely engaged in prolonged high-speed chases, and did not attack prey that were vigilant. We propose that the tiger sharks' foraging tactic is one of stealth, and sharks rely upon close approaches to prey in order to be successful. This study shows that using appropriate analysis techniques and a variety of field methods it is possible to elucidate the factors influencing habitat use and gain insights into the foraging behavior of elusive top predators.[51]

The researchers are also investigating to what extent the presence of tiger sharks modifies the behaviour of the bay's marine inhabitants on which they prey.

Orectolobiformes (Plate 7)
These are the carpet sharks, the common name deriving from their camouflage-like patterned colouring and multiple tassels (dermal lobes), which can make some resemble fabulous

underwater Persian carpets. The order of carpet sharks is tremendously varied, from the slender metre-long epaulette shark to the gigantic whale shark.

Classification

- Fourteen genera in seven families
- 31 species

Biology

- Nasal barbels
- Nasoral grooves connecting mouth to nostrils
- Five gill slits
- Two spineless dorsal fins
- Elongated upper caudal lobe
- Striking skin patterning
- Oviparous or ovoviviparous reproduction

Habitat

- Warm temperate and tropical waters
- No freshwater species
- Most common in the tropical Indo-West Pacific
- Mostly bottom dwellers
- Mostly nocturnal

Blind sharks

These small sharks use their barbels to seek out prey through electroreception, with their large spiracles assisting in breathing as they forage in shallow, particle-filled water. There are two endemic Australian species. The fairly common blind shark (*Brachaelurus waddi*) is found along the coasts of southern Queensland and New South Wales, generally in shallow waters among reefs and seagrass beds. It grows to about 1.2 metres but is generally smaller. The rare Colcloughs shark

(*Brachaelurus colcloughi*), also known as the blue-grey carpet shark, is known only from the Queensland coast and grows to just 75 centimetres. When out of water, blind sharks try to protect their eyes by closing them, hence the name.

Shy sharks

Shy sharks are endemic to the waters of South Africa. In 2006, a new species, the Natal shy shark (*Haploblepharus kistnasamyi*), was discovered off the coast of KwaZulu-Natal. The term 'discovered' is relative here, and provides a working example of how classifications develop and change. This little shark (maximum length about 50 centimetres) had previously been considered a 'colour variant' of the puffadder shy shark (*Haploblepharus edwardsii*) endemic to waters further south along the Cape coast, until South African ichthyologists Leonard Compagno and Brett Human wrote it up in the systematics journal *Zootaxa* as a new species, because of its 'stockier build, less depressed head and trunk and more depressed caudal peduncle'.[52] The colour variation is described as

> Southeastern Cape form: sandy brown with 7 reddish-brown saddles bordered by black, and numerous small, dark brown and white spots between saddles; white below. Natal form: body cream in color with darker brown saddles and irregular white spots; white below.[53]

Shy sharks eat a variety of prey including small teleosts such as anchovies, squid, crabs, shrimps and polychaetes. Their small mouths contain about 30 rows of teeth in each jaw. In turn they are preyed upon by larger sharks, seals and black-backed kelp gulls. They are also taken by recreational anglers. Like blind sharks, shy sharks protect their eyes when taken from water, but they do so by wrapping their tails across their eyes.

Wobbegongs (Plate 8)

The Aboriginal name of this flattened shark identifies it with Australian waters, although it has been found as far north as Japan. 'Wobbegong' is thought to translate either as 'living rock' or 'shaggy beard'. It is a shallow-water, bottom-dwelling predator, almost invisible in rocky, algae-covered coastal terrain. While a few species are quite small, the larger 'wobbies' grow to almost three metres, and weigh about 70 kilograms.

The wobbegong has a broad, flattened head with groups of sensitive dermal lobes protruding from the sides of the head, in front of the eyes and the lips. These help it to detect prey in murky water. Large spiracles behind the eyes enable it to breathe while motionless, and to avoid drawing in silted water. The heavy body tends to be the same width as the head, narrowing abruptly at the pelvic fins, above which is the first dorsal fin. The tail is narrow with a low caudal keel. It is not the body plan of a fast pursuit predator but the camouflaged wobbegong is perfectly suited to ambushing prey that comes within range. The wobbegong's terminal mouth and powerful protruding jaws contain wickedly sharp, long teeth. It lunges and grabs its prey, which is then sucked in by its pharyngeal muscles. The wobbegong generally waits for the prey to die, before swallowing the victim whole. It is a nocturnal feeder with a preference for octopus, teleosts and crabs. It also stalks its prey by 'walking' on its pectoral and pelvic fins.

The large spotted wobbegong (*Orectolobus maculatus*), combining shades of yellow, green and brown, is intricately patterned with leaf shapes and circles. Reports have it growing to three metres. The ornate wobbegong (*Orectolobus ornatus*) is described as 'extremely ornate and variegated; yellowish brown to grayish brown with darker, corrugated saddles;

each saddle, and the many paler bluish to whitish patches, bordered with lines of small, black spots; a prominent white spot behind each spiracle; pale ventrally'.[54]

Wobbegongs are ovoviviparous, the embryos receiving nourishment from the yolk and also secretions from the mother's uterine wall. Gestation is about eleven months, with litters varying considerably in size, from fewer than ten to more than 30. It is thought that these differences may be linked to migratory movements.[55]

In an interesting example of the fluid and dynamic nature of scientific classifications, in 2006 the banded wobbegong was successfully redescribed as being not one but two species, *Orectolobus ornatus* and *Orectolobus halei*—the smaller *ornatus* previously being classified as the juvenile *halei*. The famous Australian ichthyologist Gilbert Whitely had proposed *O. halei* as a distinct species as early as 1941, but a subsequent lack of specimens placed it in synonymy with *O. ornatus*. Charlie Huveneers successfully argued his case for a new description on this comparison:

> *Orectolobus ornatus* differs from *O. halei* in colour pattern, a smaller adult size, fewer dermal lobes at the posterior preorbital group, lower vertebral and spiral valve counts, and the absence of supraorbital knob. Morphometrically, *O. ornatus* has a longer pelvic fin to anal fin interspace, smaller pectoral fins, smaller head dimensions, and relatively smaller claspers in mature specimens.[56]

Despite being a common and easily studied shark, found in good numbers around most of Australia's enormous coastline, as recently as March 2008 two new wobbegong species were discovered, bringing the total known species to nine.

The two newly described species were at first thought to be juveniles of known species. One, the dwarf spotted wobbegong (*Orectolobus parvimaculatus*) grows to about one metre, far smaller than the two-metre spotted wobbegong.

Less rigorous on a scientific level, but no less interesting, is this observation from Julian Rocks, a popular dive spot located off Byron Bay:

> Crayfish form commensal relationships with wobbegongs. Wobbegongs eat octopus, and octopus eat crayfish. By living in close proximity to the wobbegong, the crayfish serves the wobbegong as bait and in turn receives protection.[57]

Divers interact continually with wobbegongs, which are not aggressive, but bite people more often than any other Australian shark species, perhaps because they seem docile and unsharklike, so are harassed or, being shallow-water dwellers, are trodden on. When disturbed they do, however, issue an unsubtle warning:

> The most spectacular snap I ever witnessed was at Byron Bay. Diving with local guide, Pete Murphy, at Julian Rocks, he was giving us a guided tour around Cod Hole. Swimming over a gutter we spotted a monster 3m long spotted wobbegong, with a head close to a metre wide. Pete decided he was going to pat this massive shark, so I just sat back to watch. As Pete reached out with his hand the wobby quickly turned towards him, opened its mouth wide and snapped its jaw shut just millimetres from Pete's mask. This shark was big enough to have engulfed not only Pete's head, but half of his torso. The wobby's warning snap had the desired effect, with a very pale looking Pete quickly backing away.[58]

Nurse sharks

This shark family of just two genera is widely distributed in warm shallow waters from Africa's east coast across to the central Pacific. The tawny (or tawny nurse) shark (*Nebrius ferrugineus*) reaches a maximum length of more than three metres and is a fairly distinctive shark, being uniformly tan-coloured with large, pointed dorsal fins set well back, large, sickle-shaped pectoral fins and a long upper caudal lobe. It has a large head, small eyes set back from a small underslung mouth containing small, multicuspid teeth and a pair of barbels connected to the nostrils, the latter being connected to the mouth by nasoral grooves. What is such a small feeding apparatus doing on a creature of this size? The tawny nurse has developed a method of feeding similar to that of the wobbegong. It has powerful gill and throat muscles, which it uses to suck in prey, using its pharynx as a powerful vacuum. If the barbels detect prey hiding in a crevice, the shark will attempt to hoover out the prey.

The tawny shark is a nocturnal hunter, preferring shallow waters associated with reefs, channels and sandy or seagrass bottoms. During the day, groups of nurses sleep together under ledges and overhangs. They are commonly seen like this by recreational reef divers, leading to a mistaken belief that they are harmless:

> Due to their apparently docile nature, Nurse Sharks have gained a reputation for being 'harmless'. They are not. If touched, grabbed, poked, or otherwise molested, Nurse Sharks often turn on their human tormentors with astonishing speed. Clamping on with their short, powerful jaws and multi-cusped teeth, Nurse Sharks typically roll over and over in the water—like a crocodile in a 'death roll'—causing nasty bruises and sometimes massive tissue damage. Rescu-

> ers often must carry the 'victim' out of the water with the shark still attached, the animal stubbornly refusing to let go until it suffocates or is killed. It is therefore always best to let a 'sleeping' Nurse Shark lie.[59]

The nurse shark also defends itself by reversing its inhalatory capability, that is, by ejecting powerful streams of water at the source of the threat. Not surprisingly, this has given rise to the common name 'spitting shark'. It is also called the sleepy shark, for its habit of aggregating in resting groups—stacked one atop another—in caves. The tawny shark is ovoviviparous, giving birth to small litters of fewer than ten, the newborn being an impressive 40 centimetres long.

Zebra shark (*Stegostoma fasciatum*) (Plate 9)
This species is noteworthy on a number of accounts. It is one of the few large carpet sharks, possibly growing to a maximum length of about 3.5 metres, although adults are generally about 2.5 metres. Despite its size, it is harmless to humans. Juveniles have light stripes and blotches on a dark skin which morph into dots on a paler skin in adulthood. Furthermore, while zebra shark is its common name in Australian waters elsewhere it is known as the leopard shark (*Triakis semifasciata*) (it is widespread across coastal Indo-Western Pacific). There is a ground shark commonly known as the leopard shark, but it is quite different, physiologically and behaviourally, to *S. fasciatum* and is restricted to the coastal waters of the west coast of the United States.

The zebra shark's patterning is not its only distinguishing feature. 'This distinctive species, with a very long, blade-like caudal fin, is unlikely to be confused with any other.'[60] The tail is half the length of the body but is not raised, so is more like a tapering of the body proper rather than an appendage. The zebra shark has a transverse barbeled mouth and small teeth, each comprising three sharp points. Its eyes are small and it has large spiracles. The flanks are shaped by a pair of prominent ridges running from behind the head to the caudal fin. The pectoral fins are large and wide, the underside of the body flattened. It is a body plan designed for slow movement, gliding and motionlessness. Typically, the gills pump water while the animal rests during the day. At night, it feeds on snails, crustaceans and small fish on or near the shallow bottom, and its slenderness enables it to squirm into cracks and crevices in search of prey.

Despite being common in warmer Australian waters, from the south coast of New South Wales around the Top End to the mid-coast of Western Australia, this shark, like so many others, is essentially a mystery. 'Leopard sharks aggregate annually in shallow coastal waters of southern QLD over the austral summer and, to date, there is very little information on this species in the wild and the function and duration of this aggregation is unknown.'[61] For this reason, the University of Queensland has instigated a research program into the ecology and population dynamics of leopard sharks, intended to determine likely adverse impacts on the species through fishing in the western Pacific, where it is taken commercially.

The zebra shark is oviparous. The large dark egg cases, which measure about fifteen centimetres by eight centimetres, are covered in external fibres which anchor them to rocks and the sea floor.

The whale shark (*Rhincodon typus*)

This is the planet's largest fish, growing to at least twelve metres and weighing 15 000 kilograms. (Its shape was memorably described by David Stead as being like 'a prodigious tadpole'.[62]) The whale shark is a warm-water inhabitant, its range described as 'worldwide' in tropical and warm temperate waters, but so little is known about this shark there is no real knowledge of its habitat and numbers. In all likelihood, it is rare. It was discovered early in the nineteenth century, when a specimen was washed ashore in Table Bay, Cape Town; until the 1980s there had been just a few hundred reports of the whale shark, since when intensive studies have taken place.

This huge animal has two other distinguishing features: its extraordinary skin patterning and cavernous gape. The skin's basic colouring is blue–brown above with a white belly below. The upper body is covered with distinctive vertical and horizontal stripes which create a checkerboard effect. Inside each square is a creamy white blob. The mouth has up to 3000 tiny teeth. So gigantic is the shark's maw that when it is fully opened the head seems to disappear altogether.

The first dorsal fin is large and set well back towards the large semi-lunate tail. There are prominent ridges along the flanks. When the mouth is closed the head appears to be very flat, broadening into a bulky, powerful, streamlined body built for slow cruising. Long believed to be oviparous, it has now been established that the whale shark gives birth to live young—300 were counted in a female caught in Taiwanese waters. The frequency of reproduction and rate of gestation are unknown. The five gill slits are very large and contain spongy filtering screens designed to capture the tiniest of zooplankton (warmer waters being less rich in zooplankton than temperate waters which benefit from particularly plankton-rich cold upwellings). There is nothing discriminatory about its feeding:

> The bellows-like gill pouches make the whale shark a versatile filter-feeder, enabling it to consume a wide variety of planktonic crustaceans and even small to mid-sized fishes such as sardines, anchovies, and mackerels. Because of its ability to suck food into its mouth, the whale shark is not dependent upon forward motion to operate its filters and often assumes a vertical posture when feeding. It has been reported that whale sharks enhance the efficiency of vertical feeding by 'bobbing' up and down in 15 to 20-second cycles, pausing at the surface to allow food-bearing water to rush into their mouths and strain through their spongy gill plates.[63]

Seasonal whale shark populations frequent numerous locations in the three major oceans. In Taiwanese and other Asian waters they are subject to commercial exploitation for their flesh and fins, whereas in Western Australia's Ningaloo Marine Park they are protected and studied, while also generating considerable tourist income through dive companies. A whale shark identification programme, using photographs taken of each individual's unique patterning, has generated a large database which is of considerable value in building up a life history profile of this species. Even so, much has to be learned about the largest and most docile of sharks (though it has been known to headbutt pestering boats). Why the dramatic skin patterning? Where do they breed? How long do they live? Why, alone among the carpet sharks, are they not bottom dwellers? Are they endangered?

The whale shark epitomises this about sharks: the more research is conducted on them, the more questions remain to be answered.

Heterodontiformes

Classification

- One genus in one family
- Nine species

Biology

- No nictitating membrane
- Large ridge above each eye
- Nasoral grooves connect nostrils to mouth
- Sharp front teeth, blunt crushing back teeth
- Five gill slits
- Two dorsal fins with large spines
- Anal fin
- Oviparous reproduction

Habitat

- Bottom dweller in shallow marine environments
- Warm temperate and tropical distribution
- Western and eastern Pacific Ocean, western Indian Ocean

Port Jackson shark (*Heterodontus portusjacksoni*)

Port Jackson is the body of water around which the city of Sydney is built. Soon after the First Fleet came ashore on 26 January 1788, one of the local sharks was described:

> . . . the skin is rough, and the colour, in general, brown, palest on the under parts: over the eyes on each side is a prominence, or long ridge, of about three inches; under the middle of which the eyes are placed: the teeth are very numerous, there being at least ten or eleven rows; the forward teeth are small and sharp, but as they are placed more

backward, they become more blunt and larger, and several rows are quite flat at top, forming a kind of bony palate, somewhat like that of the Wolffish; differing, however, in shape, being more inclined to square than round, which they are in that fish: the under jaw is furnished much in the same manner as the upper: the breathing holes are five in number, as is usual in the genus: on the back are two fins, and before each stands a strong spine, much as in the Prickly Hound, or Dog, fish: it has also two pectoral, and two ventral fins; but besides these, there is likewise an anal fin, placed at a middle distance between the last and the tail: the tail itself, is as it were divided, the upper part much longer than the under . . . This was taken at Port Jackson, but to what size it may usually arrive cannot be determined; perhaps not to a great one, as the teeth appear very complete.[64]

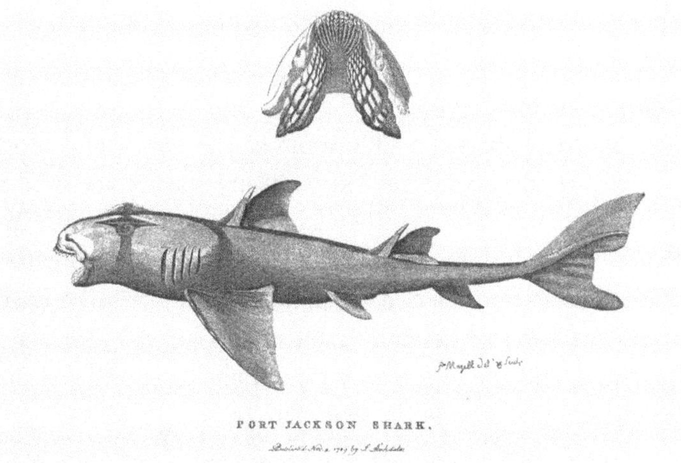

Illustration of a Port Jackson shark in The Voyage of Governor Phillip to Botany Bay, *1789.* (Libraries Board of South Australia, Facsimile Editions No. 185, 1968)

The Port Jackson shark reaches a maximum length of about 1.5 metres, although is generally smaller, and is the largest of the horn shark family, a small group of sharks also known as bullhead sharks and tabbigaws. Like so many other shark families, the horn sharks have their own characteristics and shark smarts. Their skin colouring, nocturnalism and defensive spines on each dorsal fin protect them from predators such as other sharks and seals. As described above, their front teeth are sharp, the back teeth pavement-like, ideal for grinding the shells of crustaceans. In summer adults migrate south from the east coast of mainland Australia towards the cooler waters of Tasmania. The adults mate in deep water then, as philopatric (literally, 'territory-loving') sharks, return in winter to the same sandy-floored caves along the mainland coast. They rest in groups in these caves during the day. The female wedges her large, hard-cased eggs into rock crevices, where they take about a year to hatch. Up to 100 or so young at a time can be found in shallow, protected nursery areas.

Lamniformes

The mackerel sharks include some of the fastest and most efficient of the ocean's predators. Many species are similar in body characteristics to the Carcharhiniformes, which indicates an evolutionary design for large, muscular, torpedo-shaped sharks to catch and consume large, speedy prey. Even so, the mackerel shark family also contains surprises—a 'playful' shark, and the two species that might be considered the beauty and the beast of the shark family.

Classification

- Ten genera in eight families
- Approximately 16 species

Biology

- No nictitating membrane
- Mouth extends well behind from the eyes
- Five gill slits
- Large teeth
- Two spineless dorsal fins
- Anal fin
- Large similarly sized caudal lobes
- Ovoviviparous reproduction

Habitat

- Coastal and open ocean
- Cold to tropical waters worldwide

Goblin shark (*Mitsukurina owstoni*) (Plate 10)
According to Richard Ellis, the famed US marine artist and shark expert,

> This seems to me the strangest of all the sharks. It looks like some kind of prehistoric survivor, an experiment in shark design that doesn't seem to work. And yet, by definition, it does work. *Triceratops,* the dinosaur with three horns, is long gone, as are *Pteranodon* and hundreds of other 'impossible' animals. There is little that can be said about this mysterious shark, because so little is known about it. And yet, we have the most curious, incontrovertible fact of all: *Mitsukurina* lives.[65]

The goblin shark is totally unlike other big mackerel sharks, with their classic torpedo shapes and vivid presence in the popular culture as ugly maneaters. Yet the word 'goblin' itself conjures up an image of evil ugliness. A specimen was caught by a fisherman off Japan and subsequently described in 1898 by ichthyologist (and eugenics proponent) David Starr Jordan. These sharks still appear to be most common off Japan, although their patchy distribution is worldwide in deep waters below 200 metres.

In its fundamental body plan the goblin shark is, more or less, 'sharklike', except for three strikingly unusual characteristics. The first is that it has a forwardly protruding, ledge-like snout; the second that its jaw is as mobile as a limb; the third that it is pinkish in colour, with blueish fins. It grows to almost four metres. The goblin shark's rarity means that not much is known about the species (the only known one of its kind), but its unusual colouring comes from the blood vessels which are visible through its fine, translucent skin. Its fins and flesh are soft, its wavy tail elongated, all of which suggest a slow swimmer. What is certain is that the snout, crammed with ampullae of Lorenzini, is a prey detector. Scientists, however, are rethinking the long-held assumption that the goblin shark is exclusively a bottom dweller, using its snout to seek out prey hiding in the substrate. It may also be a mid-level feeder, but whatever the case, its feeding mechanism is exceptional:

> The jaws are modified for rapid projection to aid in the capture of prey. The jaw is thrust forward by a double set of ligaments at the mandibular joints. When the jaws are retracted the ligaments are stretched and they become relaxed when the jaw is projected forward. The jaws are usually held tightly [in the head] while swimming and function like a catapult when the animal wants to feed. Its slender narrow teeth suggest it mainly feeds on soft body prey including shrimps, pelagic octopus, fish, and squid. It is also thought to feed on crabs. The posterior teeth are specialized for crushing.[66]

The goblin shark also has a large basihyal and expandable pharynx, suggesting that prey caught in its needle-sharp teeth are then sucked into the stomach. The goblin shark's oil-filled liver contributes almost 25 per cent of its body weight. This means that it has near-neutral buoyancy, allowing it to drift, waiting for prey to swim past. It is speculated that the goblin shark might also partake in nightly vertical migration, moving up and down with the fishes and cephalopods upon which it preys.

Crocodile shark (*Pseudocarcharias kamoharai*)

This metre-long oceanic shark has the distinction of being the smallest of the lamnids. It is found worldwide in tropical waters to depths of about 600 metres, with occasional specimens found inshore. The body is slender and firm with a strongly forked asymmetrical tail. Its most noticeable features are its huge eyes and elongated gills slits extending to the top of the head. Not much is known about the crocodile shark, which is thought to be a vertical migrator. The skin is brown, sometimes with white speckles.

Mako sharks (Plate 11)

Mako is the Māori word for shark. There are two known species of mako shark, the shortfin (*Isurus oxyrinchus*), and the rarer longfin (*Isurus paucus*). Superlatives abound in describing the shortfin mako, which outclasses every other shark in its combination of beauty, elegance, power and speed. At a maximum length of about four metres and a weight of some 500 kilograms, this shark's most distinguishing features are its big, pointed snout above permanently exposed multiple rows of hooked teeth, huge, glistening black eyes and brilliant two-tone colouring of indigo and white. A major pelagic species—though it does come inshore—the shortfin mako's range is truly global in warm waters, comparable to that of the blue shark.

Unfortunately, the shortfin mako is also one of the most lucrative of the elasmobranchs. It is targeted commercially for its flesh, its teeth and its jaws, and for the satisfaction of mastering the planet's ultimate fighting fish: 'It can be an especially dangerous adversary when gaffed or tail roped, and there are countless stories of violent mako encounters at boat-side!'[67]

The mako shark is supremely hydrodynamic. Behind the sharp snout and cylindrical forward body, the narrow dorsal fin is upright and prominent, the pectoral fins broad-based and tapering to pointed ends. Water flow over the tightly packed, small dermal denticles is further improved by the presence of ridged caudal keels in front of the large tail of equal-sized crescent-shaped lobes. The shortfin mako's superbly engineered tail, together with its endothermic circulatory system (see Chapter 5), gives it tremendous powers of acceleration. It can shoot through the water—and sometimes high out of it—reportedly at speeds in excess of 70 kilometres per hour.

The shortfin mako targets fast-moving prey, including tuna, bluefish, other sharks, and probably dolphins and porpoises. There is at least one known record of a swordfish rostrum being found in a shortfin mako's stomach.

Makos are solitary and migratory and may follow warm currents as part of their lifecycle, but their movements and blue-water habitat make them difficult to study. Interestingly, one American research program that tagged a mako off North Carolina in 1984 retrieved the shark nearly thirteen years later; it had been caught by a commercial fishing vessel off South Carolina, implying philopatric behaviour. The southern California bight on the west coast of the United States is known to be a shortfin mako pupping and nursery ground. The ovoviviparous young, averaging about fifteen in a litter, can exceed 70 centimetres and weigh more than ten kilograms.

Estimating an animal's lifespan is critically important in understanding the dynamics of its population. This is especially important with the shortfin mako, given the lack of knowledge about it and its vulnerability to commercial exploitation. One program designed to estimate the lifespan of shortfin mako sharks used what is known as 'bomb dating'. In the 1950s and 1960s atmospheric thermonuclear testing released enormous amounts of radiocarbon into the atmosphere. This eventually settled in the oceans, entering hard organic structures such as corals, teleost earbones and calcified elasmobranch vertebrae. Researchers are able to use these radiocarbon deposits to calculate the age of an individual organism. The California State University program employs this technique with great precision: 'The [shortfin mako vertebrae] cores will be weighed to the nearest 0.1 mg and stored in sterile plastic cryo vials in preparation for radiocarbon assay by Accelerator Mass Spectrometry (AMS) at the Center for Mass Spectrometry, Lawrence Livermore National Laboratory, Livermore, California'.[68] The program selected 54 vertebrae collected between 1950 and 1984. Results confirm that the shortfin mako lives for at least 31 years.

Porbeagle (*Lamna nasus*)

This unusual Cornish name is a portmanteau word, combining porpoise and beagle. (It was coined, however, long before Lewis Carroll invented the word 'portmanteau'.) The porbeagle is frequently mistaken for a mako shark. Furthermore, it is also often confused with its close relation the salmon shark (*Lamna ditropis*). There are discrete porbeagle populations in the western and eastern Atlantic Ocean, and in the southern hemisphere. The Atlantic porbeagle reaches a maximum size of about 3.5 metres and weight of 230 kilograms. Adults mate in autumn or winter and gestation is about nine months, with a litter of about four young live-born. And, 'the little data that exists for the Southern Hemisphere populations indicates that they may be out of phase with those of the Northern Hemisphere, giving birth off New Zealand and Australia in winter'.[69] Despite the confusion between the two, the porbeagle is stouter than the mako and has straighter teeth. It has a white splotch at the rear base of its dorsal fin and the blue of its upper body is greyer than the intense blue of the mako. The porbeagle is unique in that it has a second caudal keel, a ridge on the upper part of the lower tail lobe. Also, it is not exclusively a temperate-water shark, being found as far north as the Bering Straits in the Pacific and northern Scandinavia and Russia in the Atlantic.

L. ditropis, the salmon shark, is found only the North Pacific and there is presumably a good reason why the two species do not, apparently, co-exist in the same waters. Both prey heavily on schoolfish and are likely to have their seasonal movements dictated by their prey's migration. Unsurprisingly the principal diet of L. ditropis is Pacific salmon. Like other mackerel sharks, however, both species are opportunistic apex predators although neither is as fast as the mako, which again reduces the level of direct competition. Both the porbeagle and the salmon shark have adapted to their cooler pelagic habitats by evolving an ability to raise their body temperatures as much as fifteen degrees Celsius above the ambient water temperature.

The porbeagle displays a social intelligence not expected in a shark:

> The Porbeagle is among the very few fishes that seem to exhibit play behavior. There have been a few, sporadic accounts—principally from off the Cornish coast—of Porbeagles playing with floating objects, both man-made and natural. Porbeagles have been reported rolling while swimming along the surface, repeatedly wrapping and unwrapping their snouts and forward portion of their bodies in kelp fronds, which often trail behind the shark like rubbery streamers. Sometimes a Porbeagle thus engaged was observed being chased by other Porbeagles . . . The repetitive nature of this behavior is highly characteristic of what in other mammals would unhesitatingly be termed play. But because the Porbeagle is a 'mere' fish, ethologists are reluctant to use that term. The Porbeagle, of course, doesn't care about such semantic contention, and does what it does for its own reasons.[70]

Grey nurse shark (*Carcharias taurus*) (Plate 12)

Confusion surrounds this shark's classification, because it has a number of interchangeable, local common names and there are a number of closely related species. Thus the grey nurse shark found in Australian waters 'is also known scientifically with the synonyms *Odontaspis taurus, Eugomphodus taurus* and *Carcharias arenarius*'.[71]

The Australian grey nurse shark is currently classed as one of four species in the family Odontaspididae (although there may be two more species). Its common names in other localities include the sand shark, sand tiger shark, ground shark, slender-tooth shark, ragged-toothed shark and spotted ragged-toothed shark. The group is also referred to as the snaggletoothed sharks.

The grey nurse is a heavy-bodied mackerel shark with a somewhat humped back that can probably grow to over 3.5 metres in length and weighs in at over 150 kilograms. It possesses large, similar-sized dorsal fins, the first set well behind the pectoral fins. The pectoral and anal fins are also large, as is the heterocercal tail. Unlike the super-fast sharks, however, the grey nurse's body does not taper significantly towards the tail and the overall body plan indicates a slow mover. The upper body, covered in loosely spaced dermal denticles, is grey–brown. Juveniles have reddish or brownish spots which eventually fade. The eyes are quite small and, unlike other large, predatory sharks, lack the protective nictitating eyelid. As its common names suggest, this is a toothy family: the mouth contains about 90 long, curved pointy teeth, their multiple rows exposed in a permanent gape underneath a pointed, upturned snout.

The grey nurse and its close relatives inhabit coastal waters in fairly well-defined regions: southern Brazil; the east coast of the United States; parts of west and north-west Africa; South Africa; parts of the Mediterranean; the Red Sea; and Australia north to China. Researchers are able to be so precise because although the grey nurse migrates as part of its life cycle, it has a set home range to which it returns. Preferred habitats are deep sandy gutters and caves, and a grey nurse cave may be likened to a terrestrial social mammal's den. The grey nurse preys upon schoolfish as well as small sharks, rays and squid, and is considered to be one of the few types of shark which hunt cooperatively. Small groups of resting grey nurses have been seen drifting just above the sandy bottom, almost motionless.

The physiology and appearance of these sharks render them doubly vulnerable to human interference. On average, a mother gives birth to just one pup per year. Uterine cannibalism results in a maximum of two pups. This very low rate of reproduction heightens the adverse effects of commercial bycatch and recreational angling, and means that juvenile population numbers may be unable to sustain the natural predation of larger sharks. Worse still, for decades this docile shark's deceptively savage appearance meant it was considered to be a maneater. According to ichthyologist David Stead, throughout the first half of the twentieth century Australian newspapers inevitably reported shark attacks as being perpetrated by 'grey nurse' sharks. Well into the 1970s, the east coast grey nurse populations were targeted as a threat to swimmers. So many sharks were caught that in 1984 the New South Wales government became the first in the world to pass legislation protecting an elasmobranch, when it listed the grey nurse as a protected species. Today there are thought to be fewer than 500 individuals on the east coast, and the grey nurse is listed as Critically Endangered. The population along the coast of Western Australia is listed as Vulnerable.

Megamouth shark (*Megachasma pelagios*)

This shark was unknown to science until 1976, when one became entangled in the anchor line of a US Navy ship off Hawaii. The process of describing and classifying the specimen took seven years, but controversy remains as to its precise relationship to not only other mackerel sharks but also to the basking shark which, like the megamouth, is a filter feeder. By the end of 2006, nearly 40 specimens had been sighted or captured, the majority in open ocean waters off Japan (home to those other rarities, the frilled and goblin sharks) and the

Philippines, although a few have been recorded off the west coast of the United States and individual specimens recorded as far afield as Western Australia, South Africa, Mexico, Senegal, Brazil and Ecuador. Each megamouth is known by its chronological number, surely a unique occurrence in the animal kingdom.

The megamouth is, therefore, rightly described as the rarest known shark, and almost everything about its behaviour must be inferred. The largest specimen caught was 5.49 metres long. Another weighed over 1000 kilograms. It has small dorsal fins, a long tail to propel the body slowly through the water and large, tapering pectoral fins. The head is huge, wide and bulbous, in order to support the great terminal mouth. Each jaw contains about 50 rows of teeth, although only a few rows are functional at any one time. The inside of the mouth is silvery, the tongue dark purple. The purpose of a distinct white stripe above the mouth is not clear—perhaps it has some function in feeding, or as a distinguishing mark. The shark's skin is dark, turning black when exposed to air.

The skin itself is soft, as are the muscles and connective tissue, and the cartilage has 'poor calcification'.[72] (There is no better evidence for these attributes than to look at a photograph of a dead megamouth. Typically, the whole facial structure appears to have collapsed, the jaws themselves sagged and slewed. Despite the somewhat unappealing, 'flabby' nature of megamouth flesh, a number of the specimens caught in the Philippines were eaten.) The megamouth has protrusible jaws: it swims through clouds of plankton, shrimp and other tiny prey, sucking them in, then retracts its jaws and closes the mouth in order to expel water through its gills. A live specimen caught off the west coast of the United States was tracked for a sufficient period of time to establish that it stayed at a consistent depth of 150 metres during the day, rising to fifteen metres below the surface during the night. This indicates that the megamouth follows its prey, as krill undertake their nightly vertical migration towards the ocean's surface. Many of the observed specimens bear scars which suggest that while they are feeding, they in turn are being preyed upon by the vertically migrating cookie-cutter shark.

Thresher shark

The thresher shark is most distinguished by its tail. The length of the upcurved tail equals the length of the rest of its body, so that a thresher with a head and trunk length of three metres becomes a six-metre-long shark, although it weighs considerably less than other sharks of similar length (a thin-tail thresher has been measured at 7.6 metres and weighing 348 kilograms).

There are three species, widespread globally. The most common and largest, the thintail (*Alopias vulpinus*) has a vast range, inhabiting temperate waters worldwide. It is primarily an open-ocean shark, but is also found close to shore, possibly as part of its migratory or feeding patterns. The body shape—minus tail—is not dissimilar to that of the rotund porbeagle. The first dorsal fin is quite large; the second is tiny, its function being taken over by the enormous upper caudal lobe. The pectoral fins are fairly long and pointed at the tips. The dermal denticles are very small and closely overlapping. The tail is described as being like a scythe, narrowing to almost whiplike. It is a body plan which suggests great agility: a light yet muscular shark benefiting from the propulsion generated by such a huge caudal fin.

Underneath large eyes the thresher's mouth is quite small, as are its teeth, which are bladelike, curved, pointed, unserrated and extremely sharp. Its prey is thought to be schools of small teleosts such as mackerel and shad. Conventional wisdom is that the thresher uses its tail to herd and round up and then stun (or otherwise injure) these schoolfish, before feeding. Cooperative hunting by two or more threshers has been observed but despite its numerical abundance, and the fact that it is regularly in contact with humans through its persistence in harrying the catches of commercial fishing boats (especially mackerel), there remains some doubt over the precise functioning of the tail. There are not many recorded eyewitness accounts of the herding and stunning of prey.

Such a significant appendage could have taken the thresher into a pelagic niche not occupied by the larger, heavier and faster elasmobranch predators, namely, that it takes smaller, darting prey. A comparison can be made between the thresher and one of the planet's terrestrial predators: 'The cheetah's long muscular tail works as a rudder, stabilizing and acting as a counter balance to its body weight. This allows sudden sharp turns during high speed chases.'[73] Unlike the larger cats, the cheetah specialises in small, darting prey and it may be that the thresher's whippy tail gives it exceptional manoeuvrability. Its scientific name *vulpinus* is 'fox', an animal known not only for its bushy tail but also its considerable degree of intelligence. Perhaps this unscientific but pragmatic, empirical account of catching threshers gives some indication of their all-round energies and, possibly, fox-like cunning:

> Fighting thresher sharks can be very challenging. You have to endure pandemonium when hooking these sharks. If you catch one, you will know what I mean. It can drag you in any direction it wants with its tremendous speed and power,

> it jumps, or dives deep in the water, it never gets tired fighting with you, and even if you think you are winning, it goes around the pilings of the pier and cuts your line. Its long tail is another challenge. If the tail tangles with your line, the game is over, you are finished . . . But who cares even if you lose. Thresher sharks are one of the finest game fish to fight with . . . Thresher sharks migrate together, so when someone hooks one, you have a chance to catch it too. On the 23rd of May, 1998, we hooked 9–10 thresher sharks at Santa Monica Pier. It was almost new moon, high tide was around 8:30 a.m. Waves were calm and it was a kind of condition that we call thresher water . . . the first action, we saw a thresher splashing water with its tail . . . As soon as I saw my rod was moving, I grabbed it and waited until the fish start running . . . The fish willingly came close to the pier and splashed one time and start running again . . . I finally pulled the fish into a close range where my friends were able to gaff the fish. It was an 8 and half foot, and probably 80–100 lb shark. I thanked my fishing buddies and told them let's hook more. Threshers continued to grab our live mackerel. But, somehow, we lost every one of them except mine. Threshers won 9–1. But, we were all happy fighting with this amazing fish.[74]

The two other species are the bigeye thresher (A*lopias superciliosus*), primarily distinguished by its huge eyes, and the pelagic thresher (*A. pelagicus*), the smallest of the three species at about four metres.

Great white shark (*Carcharodon carcharias*) (Plate 13)
It is no coincidence that the explosion in scientific elasmobranch research coincided with the cultural branding in the

1970s of the great white shark as the sum of all our fears. Through its dramatic fictional distortion in print and on the silver screen, this most awe-inspiring of apex predators became a catalyst for an urgent reassessment of how we treat the living oceans.

The great white, also known as the 'white pointer' and 'white shark', has a worldwide distribution in temperate coastal waters, although it also travels into cooler northerly and southerly waters. As a migratory species which crisscrosses ocean basins, it is in no way exclusively 'coastal'. Some observers believe that the tiger shark is its apex predator equivalent in tropical waters, and that their preference for different water temperatures ensures minimal competition between the two species.

Despite the amount of research carried out over the past forty or so years, there is still much to learn about the great white's biology and behaviour. There are two reasons for this: first, great white populations have a habit of disappearing and satellite tagging as yet has not found out why; and second, because it is possibly the shark least capable of being kept in captivity. Great whites are considered rare, their numbers uncertain, although their aggregations at certain locations off the coasts of Australia, South Africa, the United States and Mexico have enabled considerable research into them. In 2006, Columba—a 3.5-metre, 500-kilogram great white—was satellite-tagged by a CSIRO team led by Australia's foremost great white shark expert, Barry Bruce. The team's research revealed that the great whites

> . . . move from one site to another, sometimes thousands of kilometres apart, usually at a speed of about 3 km/h. They can stop for periods of months in the same loca-

> tion . . . Evidence gathered so far suggests there are biological hotspots that may dictate the sharks' movements. These include seal colonies in places such as Port Lincoln, islands in the Great Australian Bight and near Esperance; schooling fish such as snapper, which frequent Spencer Gulf and Gulf St Vincent, and schools of gummy shark – also in the Bight – at certain times of the year. Scientists now believe sharks instinctively know where these hotspots are and travel directly to them using 'set corridors'.[75]

The great white is distinctively two-toned, with a grey or blue-grey upper half and white lower half. Its dermal denticles are tightly packed, so its skin is less rough than that of most sharks. The maximum length recorded is 6.4 metres, with a weight of 2500 kilograms. The great white weighs so much more than other large neoselachian predators that descriptive terms such as 'fusiform' and 'stout' need to be considered in context. The dorsal fin is broad and upright, the pectoral fins huge, the lobes of the tail almost equal in size. Aerobic ('red') muscles attached to the tail run almost the length of the body. The huge fins and tail, together with the shark's sheer muscular bulk, allow it to leap well clear of the surface as part of its feeding strategy. Enormous energy is required to propel thousands of kilograms at great speed vertically out of the density of seawater.

The great white's teeth are large, upright, serrated and triangular, ideal for disabling prey. Its large, black eyes roll back into their sockets at the moment of impact, to protect them from injury. The huge jaws lunge forward in a tremendously wide gape and inflict a crushing bite. As a great white matures its prey base becomes increasingly rich in fats, which contribute to its physical prowess. Marine mammals such as seals, sealions and walruses are a primary food source. Whale carcasses are scavenged for their blubber. Marine mammals live in temperate and cold waters; the great white's ability to regulate and raise its body temperature enables it to hunt in a range of water temperatures. Research into great white predation describes a five-stage process: detection, identification, approach, subjugation and consumption. In other words, great whites typically lurk around rocky pinniped colonies and ambush individuals as they leave or approach the colony. The great white is the only shark known to pop its head above the surface, looking for or scenting prey, but its favourite method of attack is from below. The shark's acute senses of vision and scent first detect and lock on to its selected victim. The shark stalks the prey that is swimming near the surface, then hurtles up at it, prey and predator erupting from the water in a cloud of spray. Larger prey such as elephant seals, which grow to about five metres and weigh over 2000 kilograms, tend to be bitten towards the rump, with the shark then backing off while the victim bleeds to death. Seals and sea lions are grabbed whole and held under the water until they die.

Great whites live seasonally in the vicinity of marine mammal colonies, where they appear to have semi-social interactions. They undertake solo migrations associated with breeding, although researchers have not yet been able to observe mating behaviour. Pregnant females travel to pupping grounds, where litter sizes vary from just two newborn to nearly twenty, the pups being about a metre in length.

Basking shark (*Cetorhinus maximus*) (Plate 14)

The basking shark is a megaplanktivore obtaining nutrition by obligate ram filtering in both normal and reverse diel vertical migration patterns. In other words, it's a passive filter-feeder and a bloody big fish, reaching a maximum length of about twelve metres and weighing up to 4000 kilograms. It is the only species in its genus. It is a widespread coastal-pelagic shark in temperate and Arctic waters but although the basking shark spends considerable time in inshore waters and is relatively common, not much is known about its behaviour and almost nothing about its reproduction. Nor do researchers know where this migratory shark spends its winter. It can be defined as an apex predator, but it is at the top of an extremely short food chain: phytoplankton, zooplankton, vertebrate.

While the basking shark has the lamnid characteristic of a large tail of almost equally sized lobes, this is its only physical similarity to its highly mobile predatory relatives. The mouth below the conical snout is gigantic and filled with hundreds of tiny hooked teeth. Just as distinguishing—and unique—are the five gill slits that virtually encircle the head. These large gill slits contain gill rakers which are the shark's primary means of capturing its tiny plankton prey. Its eyes are quite small, its olfactory sense highly developed. The pectoral fins are large, and the predominant skin colour is grey, with a pale underside. Like the goblin shark, the basking shark's liver weighs almost 25 per cent of its body weight, which creates near-neutral buoyancy.

The basking shark does not actively pump water into its mouth in order to feed but simply swims (slowly) mouth agape through zooplankton-rich patches of water. Researchers estimate that for this feeding method to sustain its bulk, an adult shark must strain some 2000 tonnes of water per hour during feeding. To survive, therefore, the basking shark must be able first to locate the rich patches of zooplankton, then to capture efficiently the tiny animals they contain.

The gill rakers are more than a humble 'sieve'. To ensure they are in optimal working order, they are shed and replaced throughout the animal's life. It was once thought that the basking shark shed and regrew its rakers during the winter, while it hibernated, but it is likely that the shark continues to feed in winter, albeit in much deeper waters. The mechanics of the gill rakers are as fascinating as the very idea of microscopic creatures creating and sustaining a giant:

> The incoming current is strained for planktonic food (e.g., fish eggs, copepods, cirripedes and decapod larvae) by a thousand or more gill rakers, each about 10 centimetres (4 inches) long, that are borne on the hoop-like gill arches in the wall of the gullet. When the mouth is opened, these rakers are erected by muscle strands connecting their bases with the branchial cartilages. Upon closure of the mouth, elastic fibres return them to a resting position, lying flat on the gill arches. Mucus secreted by the epithelium at the base of the rakers traps planktonic organisms and, when the mouth closes, these are probably squeezed into the mouth cavity by the collapsing rakers. Teeth are still present, but they are of very reduced size and represent simple modifications of the placoid scales present in the skin, with lateral expansions that appear to be vestigial cusps. During feeding, respiration continues simultaneously and automatically.[76]

It is thought that when these sharks gather to feed in plankton-rich waters during summer, the resulting social behaviour is oriented towards mate selection. Mate selection may also be the reason behind one of the basking shark's most unusual behaviours. This huge, slow creature is one of the few neoselachians that breaches spectacularly—a display which imbues it with both power and mystery. Tradition holds that the animal leaps from the water in order to dislodge parasites, but it is more likely that such impressive aerial displays are designed to attract mates.

Long-term studies of basking sharks off the coast of Britain, particularly off Cornwall, have also linked mate selection to feeding grounds. Basking sharks sometimes gather in very large numbers, and demonstrate a tendency to swim nose-to-tail. Some suggest that traditional tales of sea serpents may have arisen from the sight of many large dorsal fins, all in a row, playing follow-my-leader.

> Lead sharks undertook convoluted swimming paths similar to those seen in solitary, feeding basking sharks . . . Rearward sharks made identical adjustments in their swimming trajectories indicating they were following precisely the movements of the shark in front . . . Sharks engaged in following behaviour spent significant periods with their mouths closed, indicating feeding was secondary during this particular activity.[77]

The apparent simplicity of an animal passively feeding off plankton has been reconsidered in the light of studies of the feeding patterns of the basking shark:

> The copepods [zooplankton] are thought to be able to sense the position of the chaetognaths [marine worms] during the day and adapt the amplitude and direction of their DVM [diel vertical movement] accordingly . . . this suggests that the reverse DVM of sharks . . . may be attributable to reverse DVM of copepod prey as a consequence of chaetognaths being present in surface waters during the day. These observations point to an intriguing example of how a plank-

> ton invertebrate predator indirectly affects the behavior of a fish megaplanktivore.[78]

As with most other sharks, there is still much to learn about the second-largest fish. The inoffensive basking shark has never had an easy time of it, its sheer size giving it an aspect of horror. Their decomposing carcasses also diverge greatly from the living animal, adding to the unfortunate reputation of this mild-mannered shark. When dead,

> the entire gill apparatus falls away, taking with it the shark's characteristic jaws, and leaving behind only its small cranium and its exposed backbone, which have the appearance of a small head and a long neck. The triangular dorsal fin also rots away, sometimes leaving behind the rays, which can look a little like a mane—especially when the fish's skin also decays, allowing the underlying muscle fibers and connective tissue to break up into hair-like growth. Additionally, the end of the backbone only runs into the top fluke of the tail, which means that during decomposition the lower tail fluke falls off, leaving behind what looks like a long slender tail. The pectoral and sometimes the pelvic fins remain attached, but become distorted, so that they can (with a little imagination!) look like legs with feet and toes, and male sharks have a pair of leglike copulatory organs called claspers, which would yield a third pair of legs. Suddenly, the basking shark has become a hairy six-legged sea serpent![79]

Flat sharks

The superorder Batoidea comprises guitarfishes (Rhinobatiformes); wedgefishes (Rhiniformes); sawfishes (Pristiformes); skates (Rajiformes); electric rays (Torpediniformes); and stingrays (Myliobatiformes). They are collectively known as the batoids. Researchers use the terms 'Batoidea' and 'Rajomorphii' interchangeably, but the common term 'rays' is frequently used to refer to all members of the superorder. The exact nature of the relationship between the batoids and their 'true' shark cousins the distinctly flattish angel sharks, as well as to some bottom-dwelling dogfish sharks, remains uncertain. The fossil record seems to indicate that the rays are shark derivatives, a linking lineage known as the Hypnosqualea hypothesis, but some researchers argue a case of 'molecular phylogenetic evidence refuting the hypothesis of Batoidea (rays and skates) as derived sharks'.[80]

The batoids are at the heart of what has been described as a cladistic (classification) elasmobranch revolution. As a distinct group they are—even by the standards of the sharks already discussed in this book—under-researched and poorly known, so much so that it was not until 2004 that the batoids were formally assessed for the International Union for the Conservation of Nature's (IUCN's) Red List of Threatened Species. One reason for this is that most scientists and researchers tend to focus on the 'true' sharks.[81]

In 2004, the 631 identified species of batoids represented just over half the total number of species of the planet's cartilaginous fishes. Although more speciose than the sharks, the batoids do not have such bewilderingly diverse physical characteristics. Most, although all rules have exceptions, have a body that is flattened dorsoventrally, that is, from top to bottom, and a tail that is either narrow with very reduced fins (skates), or whiplike with no fins (rays). The batoids' characteristic disc-like shape is a result of the pectoral fins being greatly spread out and fused to the sides of the head. The eyes are on the top of the head, the mouth and gills underneath.

Behind the eyes are large spiracles to take in water that is then passed over the gills, which as often as not are barely above or on the seabed, or even submerged under the sand. The batoids' basic body plan—directly complementary to that of the 'true' sharks—perfects the distribution of the cartilaginous fishes into every conceivable aqueous niche: fresh, saline, estuarine, shallow, midwater, deep, warm, temperate, cold, clear, murky, dark or lightless.

The batoids' jaws have evolved to take every possible advantage of their life as bottom feeders. Hard-shelled prey is a major component of their diet, but this means they have to separate out shell and other indigestible or useless matter. Unlike bottom-dwelling suction-feeding sharks they do not repeatedly take in and spit out prey, a feeding mechanism which eliminates unwanted matter before swallowing but which risks exposing valuable food to scavengers. Instead, the batoids' elaborate lower jaw apparatus allows them to separate out wanted and unwanted material internally, with the latter then ejected. This is described as a 'largely hydrodynamic form of manipulation, permitting exploitation of benthic invertebrates with exoskeletons',[82] through a number of cranial muscles—which sharks lack—that manipulate the lower jaw. This is seen as the benthic equivalent of large predatory sharks being able to protrude their upper jaws when biting prey. (Elasmobranch jaw manoeuvrability is the evolutionary equivalent of teleosts using their pharyngeal jaws and tetrapods their limbs to manipulate and process food.)

Guitarfishes (Plate 15)

Classification

- Four genera in three families
- At least 47 species

Biology

- Elongated snout, tapering sharklike body with two dorsal fins
- Ovoviviparous

Habitat

- Temperate and tropical inshore waters globally

As an order they are the Rhinobatiformes, but as a suborder of the rays they are the Rhinobatoidei. Furthermore, the single-family order Rhiniformes—the sharkfin guitarfish, also known as the wedgefish or shovelnose ray—simply confounds what is already a confusing taxonomy. The Rhinobatiformes include thornbacks, fanrays, panrays, banjo rays and fiddler rays (*pesce violino* in Italian). The common names are the key here, for these fishes combine the disc and the fusiform body plans. The rhiniform rays, with their pectoral fins fused to their heads, are described as having 'sharklike dorsal fins' with a 'trunk thick and sharklike'.[83] Physiologically, if not zoologically, they appear to be an intermediate form, and there is no reason why there should not be such forms. Shovelnose rays can also be distinguished from guitarfishes by the relative placement and size of their pelvic fins and lower caudal lobes, while fiddler rays differ from guitarfishes in their snout shapes.[84]

Most guitarfishes are small or medium-sized inhabitants of shallow inshore warm and tropical waters, including estuarine environments. A few species are found in temperate waters as far south as Australia's Bass Strait. The paired, prominent dorsal fins are set well back along the thick, tapering tail, which provides vigorous sharklike locomotion, while the spread pectoral fins are effective in both acceleration and manoeuvrability. Guitarfishes swim at a positive angle: the head is slightly higher than the tail. During the day, guitarfishes tend to bury themselves in sand or other substrate, leaving just their eyes and spiracles above the surface. They are ambush predators, grabbing unwary crustaceans and bivalves, often with the help of the rostrum which, like a limb, pins the prey down and prevents it from escaping. Guitarfish have small, numerous and blunt teeth, designed for crushing shells. By night, guitarfishes are more active, actively hunting small fish. Guitarfishes are viviparous, bearing up to twelve live young.

The difficulties in classifying the batoids can be demonstrated by the case of two species of shovelnose rays inhabiting the continental shelf off the coast of north-western Australia. Theirs is an intriguing story. In the 1970s, commercial Taiwanese trawl fishing in the area led Australian authorities to set up a research program to identify exactly what was being caught in the trawlers' nets. This CSIRO program ran for ten years, and was the first thorough study of the local fish fauna. Many new species were discovered, including two small shovelnose rays which did not quite conform to known shovelnose species. As with many taxonomical orphans, for some years the literature referred to them as 'undescribed species'.[85]

The two rays, both just over 50 centimetres long, were finally described in 2004 by senior CSIRO scientist Peter Last.[86] They are the goldeneye shovelnose ray (*Rhinobatos sainsburyi*) and the spotted shovelnose ray (*Aptychotrema timorensis*), the latter described from just a single specimen taken at a depth of over 100 metres in the Timor Sea. The spotted shovelnose appears to be the northern Australian equivalent of the continent's western shovelnose ray (*Aptychotrema vincentiana*) and the eastern shovelnose ray (*Aptychotrema rostrat*), but can be distinguished physiologically by having definite white spots rather than blotches or no patterning at all on its skin; a slightly different snout apex; more angular dorsal-fin apices; a lack of dark marking at the snout's ventral apex; and fewer vertebral centra, resulting in a smaller caudal fin. Easy!

The morphometric table for the holotypes of these two new species painstakingly measures, to parts of millimetres, no fewer than 62 external parts of their little bodies which, upon reflection, are surely as worthy of admiration as those of their breaching great white cousins. As has been suggested of the shovelnose guitarfish in faraway California: 'This ancient ray has been playing it flat for over 100 million years'.[87]

Wedgefishes

Classification

- Two genera in one family
- Five species

Biology

- Intermediate body form between shark and ray
- Small blunt teeth
- Two large dorsal fins
- Well-developed caudal fin
- Ovoviviparous

Habitat

- Shallow coastal tropical and subtropical waters

The bowmouth guitarfish (*Rhina anclyostoma*) is relatively common in warm reef waters from northern Australia through the Indonesian archipelago to eastern China. It is described as:

> an unmistakable guitarfish with a broad, rounded snout, large, high pectoral fins, and heavy ridges of spiky thorns over the eyes and on the back and shoulders; jaws with heavily ridged, crushing teeth in undulating rows . . . Grey or brownish above, white below; numerous white spots dorsally on fins, body and tail; black spots on head and shoulders but no eyespots or ocelli . . . Inhabits coastal areas and on coral reefs, close inshore . . . Found on sand and mud bottoms . . . Sometimes found in the water column . . . Feeds mainly on bottom crustaceans and mollusks . . . A row of large spines present above the eye, on the center of the nape, and on the shoulder have a defensive function (can be used for butting). Caught commonly by demersal tangle net, and occasionally trawl and longline fisheries . . . Difficult to handle and can damage the catch when caught in trawls . . . Utilized fresh and dried-salted; the pectoral fins are the only part which is eaten.[88]

Sawfishes (Plate 16)

Classification

- Two genera in one family
- Seven species

Biology

- Elongated snout is a toothed rostrum without barbels
- Sharklike body with two dorsal fins
- Ovoviviparous

Habitat

- Coastal marine waters, one freshwater species

The sawfish is a ray. Its pectoral fins are distinctly detached from its head, it has large spiracles behind its eyes to take in water and its mouth, nostrils and gills are ventral. The trunk is flattened dorsoventrally but, despite this, its body form is decidedly unraylike, having a thick shark lower body and tail supported by large dorsal and pelvic fins. Its snout extends to a toothed rostrum which in some species is long enough to class the sawfish as 'gigantic', possibly exceeding both the great white and the tiger shark in total length. The scientific literature invariably describes the sawfish as a highly modified ray and it is not to be confused with the sawshark. Sawfishes are inshore dwellers in warm waters, both salt and fresh. They are among the most endangered of all marine life.

Sawfishes are classified in the order Pristiformes, comprising just two genera of approximately seven species:

> However, due to considerable taxonomic confusion this number may in fact vary between four and ten. Among the reasons for this taxonomic disarray is that many of the original species descriptions were extremely abbreviated, and in some cases not even based on specimens, or based only on isolated anatomical parts; only two of the six type specimens are available for examination today; poor representation of specimens in collections, which mostly consist of dried rostra or very young specimens; and scarcity of these animals in their natural habitat due to overfishing.[89]

(Bizarrely, the caudal fin of a sawfish that Dutch ichthyologist Dr P. Bleekers had used to describe the holotype of a specimen in 1852 was later found stashed in the body cavity of another specimen in his collection.)

The species described here are:

- Dwarf sawfish (*Pristis clavata*)
- Freshwater sawfish (*Pristis microdon*)

- Smalltooth sawfish (*Pristis pectinata*)
- Largetooth sawfish (*Pristis perotteti*)
- Green sawfish (*Pristis zijsron*)

The species that prefer freshwater environments have broad, tapering rostrums; those with a preference for saltwater have non-tapering rostrums. The evenly spaced rostral teeth are modified denticles and number from about fourteen to 23 on each side. There is not always the same number of teeth on each side. The seventh species has a distinctly narrower saw with flatter teeth. The teeth of the rostrum are self-sharpening, through constant abrasion against sand and stone, and a broken tooth will grow again as long as the base is not damaged.

No species is found exclusively in freshwater environments. The dwarf sawfish, which inhabits the shallow estuarine and tidal mangrove waters of tropical Australia, moves into freshwater rivers during the wet season, possibly in order to breed, as does the much larger and more widespread freshwater sawfish which grows to about six or seven metres. Juveniles spend their first three or four years in rivers before moving to shallow coastal waters. However, adults are also known to travel as far as 100 kilometres upriver. In the Top End of Australia, this migration may be associated with wet season flooding.

Sawfishes are recorded in river systems throughout southeast Asia, India and eastern Africa, many of which are also home to the bull shark. Given their very different feeding strategies, competition between them is likely to be minimal. In the western Atlantic the smalltooth sawfish and the rare largetooth sawfish exhibit similar habitat diversity:

> Florida's sawfish are most often found within a mile of land such as in estuaries, river mouths, bays, or inlets. They occur

> in a wide range of habitat types including grass flats, mud bottoms, along oyster bars, sand bottoms, artificial reefs, under or adjacent to mangrove shorelines, associated with docks, bridges or piers. They can also be found miles up rivers in low salinity conditions. Large sawfish can occasionally be found living in close association with artificial reefs or wrecks, hard bottoms, or mud bottoms.[90]

The sawfish's rostrum is clearly its principal evolutionary feature. How and why did such a long, inflexible toothed extension to the snout come about? Until more is known about the sawfish's biology and behaviour this question will be difficult to answer, but a chance capture of a sawfish off the coast of Queensland may provide a clue. A 3.6-metre green sawfish became entangled in a commercial gillnet and was set free unharmed, after scientists had fitted it with an acoustic tag—the first sawfish to be monitored this way in Australian waters. The tag tracked its movements over a 27-hour period, during which time it moved north along the Gulf of Carpentaria from the Port Musgrave estuary. It began feeding within an hour of its release and spent most of the tag period in water less than two metres deep. This suggests a specific design of the saw to be most effective in very shallow water for feeding, despite the fish itself growing to a hefty five metres and more.

Studies indicate that males have longer saws than females and more teeth, which could mean that the saw has a role in courtship competition and mate selection. Juvenile and small adult sawfishes use their saws to dig for and scuff up bottom prey, while larger sawfishes use them to disable prey, particularly shoaling fish such as mullet and herring. These small fishes invariably inhabit shallow waters, and shallow-water substrate provides a rich concentration of molluscs and bivalves. In other words, the prey bases of both young and adult sawfishes are located in similar waters so there is no real need for them to penetrate deeper ocean waters. Perhaps this explains why sawfishes, as coastal dwellers, have adapted to weakly saline and then freshwater systems.

Skates

Classification

- 26 genera in three families
- More than 260 species

Biology

- Pointed snout, broad pectoral disc, small narrow tail
- Paired, single or no dorsal fins
- Large dermal bucklers
- Oviparous

Habitat

- Mostly deep offshore

Skates belong to the large order Rajiformes, comprising three families divided into some 25 genera, totalling approximately 250 species, many of which are physiologically very similar and almost none of which, despite being plentiful in certain parts of the world, are well understood. Skates are the least-studied elasmobranchs. They are unobtrusive bottom dwellers, usually found in deep waters and, with the exception of some species, are not of great commercial value. Skates do not have the appeal or drama of some of their fusiform cousins but this makes them no less important in the marine biota. In fact, as the most speciose benthic elasmobranchs, they are a critical link in the food chain.

A skate is readily identifiable by its flat, pan shape and pointy snout (which has sensory capabilities), appendage-like pelvic fins, large paired claspers on males and short, straight, slender tail with miniature dorsal fins near the tip. Most species are small, some no bigger than a sideplate. Skates' dermal denticles are unique in that they are not uniform, but grow in a wide variety of shapes and areas on the body. The longer denticles are known as thorns: those along the centre of the skate's back and along the top of its tail are called bucklers; those around the eyes malar spines; and those at the disc edges alar spines. The thorns have a protective function, and it is thought that the alar spines may also help the male to grip the female during copulation. Scientific analysis of very similar dermal covering in different skate species has led to questions about the validity of some species classifications, and ongoing research is being carried out as Project Odontobase at the Museum of Natural History in Paris.[91]

Skate disc shapes vary and have various formal descriptors: circular, rhomboidal, oval, heart-shaped, quadrangular, subcircular, wedge-shaped. Disc shape, thorns and snout shape have traditionally been the three principal species identifiers. A skate swims by undulating its pectoral fins, using its tail as a kind of rudder. The pelvic fins move the animal forward with punting-like motions.

To support such wide pectoral fins, parts of the cartilage are fused to the vertebral column. Between the cranium and shoulder girdle the skate's vertebrae are fused into a tube called the synarchial. The disc is further supported by the trunk. Another feature of the adaptive flattening of the skate is its upper jaw, which has no attachment to the cranium and is thus more mobile in feeding at substrate level. The skate's mouth is surrounded by ampullae of Lorenzini, which enable the animal to detect prey beneath the sand or mud. A ventral lateral line is able to detect the small jets of water emitted by bivalves. The skate also has electrical facilities on its tail, but these are thought to be used only in communication, particularly prior to mating. Skates are oviparous and the eggs laid may take up to a year to hatch.

The thorny skate (*Amblyraja radiata*) is well-named:

> As females mature sexually, they become increasingly thornier with increasing prickles between the thorns. Males, on the other hand, lose most of the thorns on the pectoral fins except along the anterior margins. Typically for adult males, the outer corners of the pectoral fins have two rows of hooked, erectile thorns.[92]

This skate inhabits cool waters in the western and eastern Atlantic Ocean, as far north as Greenland and Scandinavia. The similarly named thornback skate (*Raja lemprieri*), common in the temperate waters of south-eastern Australia and Tasmania, is no less well-equipped:

> Orbital ridge with 6–10 thorns (forming a rosette in juveniles); median disc thorns commencing just in advance of nuchal area, usually extending along disc midline and onto tail; media thorns with 2–6 additional midlateral rows extending posteriorly . . . Dorsal surface of disc and tail with fine granulations (very dense in adults); alar thorns retractable; malar thorns well developed . . .[93]

The boreal skate (*Raja hyperborean*) has probably the widest range of the skates, being found in deep cold waters off southern Australia, New Zealand, northern Japan, the east coast of the United States, the tip of southern Africa and north of the English Channel to Greenland. It is also known as the Arctic skate; catch studies reveal large individuals being taken at depths of about 700 metres in the Norwegian Sea.[94] Its far south equivalent is the Antarctic skate (*Bathyraja eatonii*). There are far fewer species of shark and ray in stable temperature Antarctic waters than in seasonal Arctic waters.

What, then, do skates—slow movers designed for the seafloor—eat? Their diet is surprisingly varied. The large barndoor skate (*Dipturus laevis*), confined to the east coasts of the United States and Canada, preys upon

> . . . polychaetes, gastropods, bivalve mollusks, squids, crustaceans and fishes. Small individuals subsist on benthic invertebrates such as polychaetes, copepods, amphipods, isopods, crangon shrimp, and euphausiids. Larger specimens are capable of capturing larger and more active prey, including razor clams, large gastropods, squids, cancer crabs, spider crabs, lobsters and fishes . . . the thorns on the snout of barndoor skates are worn smooth, as though the snout is used to dig in the mud or sand to obtain bivalve mollusks.[95]

The Melbourne skate (*Raja whitleyi*) reaches up to two metres in length and is found in number from south-western Western Australia to the mid-north coast of New South Wales. It is Australia's largest skate, weighing up to 50 kilograms. Never to be outdone in the naming stakes, however, is New South Wales. The Sydney skate (*Raja australis*), found along the New South Wales coast, grows to a comparatively small 50 centimetres. Its short, thickish tail sprouts between three and five rows of large, sharp thorns.

Electric rays

Classification

- Eleven genera in four families
- At least 56 species

Biology

- Platelike pectoral disc, thick-bodied
- Large paired electric organs behind small eyes
- Thick tail
- Ovoviviparous

Habitat

- Temperate and tropical waters worldwide

The electric rays (order Torpediniformes) are grouped in four families: numbfishes, sleeper rays, torpedo rays and coffin rays. The Latin word *torpere* means 'to stiffen or paralyse', which effectively sums up these rays' principal weapon, their electricity-generating organs. While they bear a superficial resemblance to skates, their discs (incorporating the snout) are thicker and rounder and their tails short and broad with large dorsal and caudal fins. Extending well back and laterally behind the small eyes are the two electric organs with which the ray stuns its prey. Situated beneath the skin, through which they are visible in some species, they give the disc its bulk. The teeth are quite small. Electric rays are exclusively marine, preferring the temperate and tropical waters of the three major oceans. Their habitats range from shallow reef waters to continental shelves of hundreds of metres.

The coffin ray (*Hypnos monopterygium*), so-named because in death it bloats into a coffin shape, is endemic to a broad band of inshore Australian waters, from tropical Western Australia to the Spencer Gulf and along the New South Wales coast to southern Queensland. It is small, seldom exceeding 50 centimetres in length, with an almost circular disc. Its pelvic fins are themselves joined to form a second disc, resulting in a miniature tail. It is a muddy brown colour with loose skin lacking either dermal denticles or thorns, a characteristic of the electric rays. The eyes are tiny and set on retractable stalks relatively well back from the straight front of the disc. These stalks can assume a vertical position, to increase the animal's field of vision.

The coffin ray is relatively common in intertidal zones, although it also inhabits rocky and coral reefs. By day, it camouflages itself in sand or mud or grassy bottoms. As a bottom feeder its usual prey includes molluscs, crustaceans and bony fishes, but it seems capable of ingesting larger prey. Although a slow swimmer, it can move surprisingly quickly and its mouth can open widely. A dead coffin ray found on Umina Beach on the central New South Wales coast in 2004 had the remains of a rat sticking out of its mouth, which led to correspondence with the Australian Museum: how could such a small ray, only 45 centimetres long, take on a whole rat? Setting aside the separate puzzle of how the animal had come across a land mammal (unless it was the water rat, *Hydromys chrysogaster*), aquarium official Craig Henderson came up with empirical evidence of its ability to take large prey:

> As Unit Supervisor of Taronga Zoo Aquarium until its closure in 1992 I undertook a captive management study of *Hypnos monopterygium* as the Aquarium had never previously been successful in keeping them in captivity. We found

> that the rays would only feed on fish (only whole fish initially) that were presented to them on a plastic stick which was 'swum' past their nose. The rays required a soft substrate such as sand or fine shellgrit to bury into. Their attack of the prey was incredibly fast, without any warning, and was always accompanied by an electrical discharge. The prey item (including very large fish) would be swallowed head first, whole, very quickly before the animal would re-bury itself.[96]

The torpedo rays are the largest of the four ray families, in some cases reaching two metres. The large disc is uniformly oval, although the tail can be either long or short. Torpedo rays are found in all three major oceans. The short-tail torpedo ray (*Torpedo macneilli*) is endemic to Australian waters and common along the island's coasts excepting the Top End. Its bulging electric organs can emit a charge of up to 200 volts:

> There are several reports of fishermen being thrown several feet after unwittingly touching large specimens of this ray. The electric organs consist of a battery of hexagonal cells (sometimes visible through the skin in smaller rays) that form a honeycomb structure on each side of the body. Each cell is filled with a jelly-like fluid and is connected to an elaborate network. When fully charged, the organs operate much like a battery, with the upper surface having a positive charge and the lower surface being negative.[97]

Descriptions of the two remaining families, the numbfishes and sleeper rays, is difficult, as this 2007 IUCN report demonstrates:

> Aspects of the general biology of the numbfishes are poorly-known. The family, like other electric rays is yolk-sac viviparous, but details on the reproductive cycle, fecundity and maturity, as well as age and growth information is generally lacking . . . The Taiwanese blind numbfish *B. yangi* is known only from a limited number of specimens detailed in its original description while the dark blindar *B. moresbyi* is known only from five specimens . . . There is minimal information on the biology of the sleeper rays. The biology of the rare Natal sleeper ray *Heteronarce garmani* is 'virtually unknown' . . . The two endemic New Zealand sleeper rays of the genus *Typhlonarke* are also rare and poorly-known, although they reportedly have up to 11 pups per litter . . .[98]

Stingrays

Classification

- At least 25 genera in nine families
- At least 189 species

Biology

- Pectoral disc triangular, oval, rhomboid, often winged
- Usually slender tail with one or more serrated stinging spines
- Six gills in some species
- Ovoviviparous

Habitat

- Temperate and tropical waters worldwide
- Marine, estuarine and freshwater

The order Myliobatiformes consists of three genera, stingray, stingaree and ray, grouped in nine families totalling nearly 100 species: stingarees; giant stingarees; sixgill stingrays; river stingrays; whiptail stingrays; butterfly rays; eagle rays; cownose rays; and devil rays. They range in size from small to very large, and are found worldwide in warm and temperate waters, inshore and offshore as well as in freshwater environments. This diversity of species types and habitat is a good indicator of a successful, adaptable life form. Disc shapes vary considerably. Stingray tails are generally long and very thin (with the exception of stingarees); but almost all have toxic, replaceable stinging spines embedded in the upper section of the tail. They are mostly bottom dwellers.

The stingaree's tail is moderately thick and short and it is the only stingray to have a caudal fin. That being said, there are at least 35 species of stingaree; a profusion which perplexes rather than assists researchers. Stingarees have a 'peculiar and disjunct' habitat distribution: 'very diverse' in the eastern Indian Ocean but not known in the western or northern parts of that ocean; very diverse in the western Pacific off Australia and the eastern Pacific to Chile and California, but not known in the Central Pacific; and scattered species found in waters off Japan and Korea.[99]

Stingarees are found inshore but also at depths of up to 700 metres on continental slopes. They hunt in a manner known as hydraulic mining, in which they emit a powerful jet of water into the seabed to uncover prey. They also fan their pectoral fins to blow away sand. Stingarees not only bury themselves during the day, they also have a considerable array of camouflage, as evidenced by some of their common names: striped, sandyback, banded, patchwork, spotted, brown, yellowback and greenback. This is as much to avoid being preyed upon as to ambush prey. This shy batoid will use its stinging spine to defend itself only as a last resort. A detailed study of stingaree tail spines (from 423 individuals representing 46 Pacific Ocean species) concluded that the spine lengths and their number of serrations related to specific habitat—open water, midwater or benthic.[100] Perhaps, then, the spine plays a more complex role in the stingaree's world than is commonly assumed.

There are many species of stingray, the largest of which is the short-tail stingray (*Dasyatis brevicaudata*) which is common in waters at depths of 100 metres, especially off southern Africa, Australia below the tropics and New Zealand. Its thick, massive disc can exceed two metres in width. The body length is at least four metres, and it can weigh in excess of 350 kilograms.

The pelagic stingray (*Pteroplatytrygon violacea*) is unusual in being an open ocean dweller, as well as frequenting inshore waters. It is virtually the only ray that doesn't bury itself in sand but cruises the ocean's upper layer preying on shoalfish such as herring and mackerel, squid, jellyfish and also small crustaceans. Not surprisingly it is a regular bycatch of pelagic fisheries.

The white-blotched river stingray (*Potamotrygon leopoldi*) is one of at least 22 species of freshwater stingray. It has a very rounded disc, covered in large, distinctive white blotches. These exclusively freshwater elasmobranchs are found only in tropical rivers of South America, particularly the Amazon River, where there are frequent reports of humans being stung, occasionally causing death. Biologically, their adaptation to freshwater environments has rendered them incapable of tolerating saline water; they don't have urea in their bloodstream, nor a specialised rectal gland to expel salt. It's not known how they came to evolve in this way, but one theory is that they derive from a saltwater Pacific Ocean species trapped by the creation of the Panama isthmus. Unfortunately, their brilliantly varied colouring means that *P. leopoldi* and its close relatives are heavily targeted for the worldwide aquarium trade.

The browny-pink, plump sixgill stingray (*Hexatrygon bickelli*) may be unique. Common in the waters off South Africa, it is also found in the western Pacific. Its gills have large gill filaments and its spiracles have external rather than internal valve flaps. Another unusual feature is its elongated, translucent snout, measuring about one third of the total body length, which may have a function in rooting around for prey. The young develop inside the mother, nourished by uterine milk and their birth, too, may be unique: 'According to one investigator, a young ray is rolled up like a cigar during birth, which . . . facilitates the birth of such proportionally large young. The young ray then unrolls and swims away'.[101]

The cownose ray (*Rhinoptera bonasus*) is a fairly large species which inhabits the open ocean as well as inshore waters. It is particularly known for its long group migrations which are probably temperature-related and may be guided by a form of solar navigation. Its common name arises from the large indent at the front of its skull. The rear end of its tan-coloured disc is also snub-shaped, while the disc tapers to points, giving it a rather boxy shape. When hundreds of these cownose rays are seen migrating in clear waters, they look like underwater kites. They are found off northwest Africa, northeast South America and from the Gulf of Mexico along the east coast of the United States. Their feeding method is intriguing:

> Cownose rays use a very specific mechanism to obtain deep-burrowing prey. They locate food on the bottom substrate (benthos) through mechano- or electroreceptive detection. Once they suspect prey is there, they employ a combination of stirring motions of the pectorals while sucking/venting both water and sediment out through the gills and away from the area to create a central steep-sided cavity depression. The continued movement of the pectoral fins aids in dispersing the sediments released from the gills and increases the depth of the depression. Eventually, the food is seized and drawn into the mouth. Common prey items include nekton, zoobenthos, finfish, benthos crustaceans, mollusks, bony fish, crabs, lobsters, bivalves, and gastropods.[102]

The huge and majestic manta ray (*Manta birostris*) occurs globally in warm waters. As befits this underwater albatross, the manta's great wings, measured as disc width, attain nine metres. Their power to propel it through the water aids its filter feeding. An adult can weigh over 1300 kilograms. Size apart, the manta's most distinguishing features are its cephalic lobes, which extend forward of the head. These are malleable and act to herd plankton into the anterior mouth. Much of manta behaviour remains to be properly understood:

> Three types of jumps have been observed, forward jumps landing head first, forward jumps landing tail first, and somersaulting. Groups of these animals have been seen participating in this behavior, breaching one after the other. While it is not understood why this behavior is exhibited, some speculate it may play a role in attracting mates or is a form of play.[103]

Chimaeriformes

One of the two subclasses of the cartilaginous fishes, the chimaeras can be traced directly back to the experimental cartilaginous fishes of the Paleozoic Era and are, with some justification, described as 'hybrids' of sharks and teleosts, a kind of non-missing link.

A chimaera has some elasmobranch characteristics, namely, a cartilaginous skeleton, claspers and a poisonous dorsal fin spine, but it does not have dermal denticles or spiracles. Its eggs are fertilised internally and laid in large, hard eggcases on the seabed. Its teleost characteristics are an operculum (gill covering), a jaw fused to its skull, and combined anal and urogenital openings. Three chimaera species, the ghostshark, the Pacific spookfish and the elephant fish, are found in Australasian waters.

Spotted ratfish (*Hydrolagus colliei*)

The spotted ratfish reaches nearly one metre in length and has an eastern Pacific Ocean range from southern California to far north Alaska. It lives and feeds at depths of from 30 to 1000 metres, generally in rocky and muddy habitats. The spotted ratfish has a large head described as rabbit-like and big, shiny, emerald-green eyes. Its brown, slimy body tapers markedly to a long tail and is dappled with white spots. It uses its large pectoral fins like wings to 'fly' unhurriedly though the water, sometimes doing barrel rolls and corkscrews. The skin is smooth with no scales or denticles. Prey includes small crabs, worms, benthic fishes, and a cannibalistic tendency—all eaten with a modest set of teeth: just two in the lower jaw and two pairs of grinding plates in the upper jaw. Reproduction is unusual:

> Following courtship rituals, the female releases a spoon-shaped egg case every 10–14 days over a period of several months. Extrusion of the egg case from the female takes 18–30 hours after which the egg case hangs freely in the water, suspended from the female by a long slender extension of the egg case called the elastic capsular filament. Each egg case eventually becomes attached to the seabed or bottom sediments by the thin tendrils of the egg case. The incubation period within the egg case is approximately 12 months.[104]

Elephant fish (*Callorhinchus milii*) (Plate 18)

The elephant fish is the most anatomically distinctive of the chimaeras. Growing to just over a metre in length, it has a long, flexible, forward-protruding snout appendage covered in the ampullae of Lorenzini, with which it detects prey submerged beneath the seabed. It is found in temperate waters off South America, southern Africa and Australasia. When cruising in sunlit waters, its skin shines like aluminium foil. Elephant fish mate and lay their eggs in shallow, coastal waters, and this is when they are most vulnerable. They are harvested commercially in New Zealand and they are a favourite target of southern Australian anglers:

> Elephant fish take baits like pilchard, fish fillets and fresh squid very well and are best hooked by leaving the rod in the rod holder with enough drag on the clutch of the reel to pull the rod into a good working curve to set the hook. Anglers who hold their rod, or pounce on their rod and strike at the first sign of a bite, do miss a lot of fish. This is because elephants dislike having the bait pulled away from them before they have had a chance to eat it. Elephant fish respond

well to berley and can sometimes be enticed into feeding frenzy. Should the tide be running—as is usually the case in the Barwon River [Victoria] estuary—the berley should be in a container right down on the bottom and consist of minced up fish, pilchards or the like.[105]

NOTES

Introduction

1 Serena, F., *et al.* (2020), 'Species diversity, taxonomy and distribution of Chondrichthyes in the Mediterranean and Black Sea', *The European Zoological Journal*, vol. 87, 2020 – Issue 1, p. 497. (https://www.tandfonline.com/doi/full/10.1080/24750263.2020.1805518, accessed 9 August 2023)

2 Shark Finning - Humane Society International (hsi.org), accessed 9 August 2023.

3 https://www.msc.org/en-au/what-we-are-doing/our-collective-impact/what-is-a-fishery/ending-shark-finning?gad=1&gclid=EAIaIQobChMI2MrJ8L7OgAMVBEYrCh3e5gyyEAAYAiAAEgInrfD_BwE, accessed 9 August 2023.

4 Towers, Lucy, 'State of the Global Market for Shark Products', The Fish Site (Cork, Ireland), 14 September 2015, https://thefishsite.com/articles/state-of-the-global-market-for-shark-products, accessed 12 August 2023.

5 Black, Chris, (2008), *White Pointer South: The Tasmanian white shark chronicles*, Hobart: Ragged Tooth Productions, p. 164.

6 Serena, F., *et al.*, op. cit.

1. Shark Attack: *Controversy, Reality, Response*

1 Quirke, Antonia (2002), Jaws [BFI Modern Classics series], London: British Film Institute, p. 52.

2 Pemberton, David, personal communication with the author, 23 September 2008.

3 Capuzzo, Michael (2001), Close to Shore: The extraordinary true story of the New Jersey great white shark attacks of 1916, London: Headline, p. 174.

4 <news.nationalgeographic.com/news/2004/01/0123_040123_tvgreatwhiteshark.html>, accessed 20 August 2007.

5 www.sharkresearchcommittee.com/unprovoked_surfer.htm, accessed 12 July 2008.

6 www.msnbc.com/id/24313314/, accessed 27 April 2008.

7 http://www.flmnh.ufl.edu/fish/sharks/statistics/2007attack summary, accessed 28 April 2008.

8 https://www.floridamuseum.ufl.edu/shark-attacks/yearly-world-wide-summary/, accessed 9 August 2023.

9 Coad, Brian W. and Papahn, Frough (1988), 'Shark attacks in the rivers of southern Iran', Environmental Biology of Fishes, vol. 23, no. 1–2, pp. 131–4.

10 http://www.dogexpert.com/FatalDogAttacks/fataldogattacks.html;http://www.flmnh.ufl.edu/fish/Sharks/statistics/statsw.htm, accessed 6 December 2008.

11 https://www.floridamuseum.ufl.edu/wp-content/uploads/sites/82/2019/01/ISAF_Questionnaire_updated-2019.pdf, accessed 9 August 2023.

12 Steel, Rodney (1992), Sharks of the World, London: Blandford Press, p. 52.

13 http://www.taronga.org.au/tcsa/conservation-programs/australian-shark-attack-file/general-information.aspx

14 http://www.surfersvillage.com/surfing/13317/news.htm>, accessed 24 August 2007.

15 http://www.news.com.au/story0,10117,17759065-1248,00.html, accessed 24 August 2007.

16 Ibid.

17 http://www.abc.net.au/7.30/content/2006/s1544350.htm, accessed 24 August 2007.

18 http://www.news.com.au/story0,10117,17759065-1248,00.html, accessed 24 August 2007.

19 http://www.abc.net.au/7.30/content/2006/s1544350.htm, accessed 24 August 2007.

20 *TheAustralian*,25January2007,p.7;TheMercury,25January2007, p. 3; <www.smh.com.au/news/national/diver-savedfrom-jaws-of-a-shark/2007/01/23/1169518709539.html?page=fullpage#>; <www.inews.independent.co.uk/environment/article2190042.ece>, accessed 1 March 2007.

21 *The Australian*, op. cit.

22 http://www.smh.com.au/news/national/diver-saved-from-jaws-

of-a-shark/2007/01/23/1169518709539.html?page=fullpage#, accessed 1 March 2007.

23 *The Mercury*, 12 May 2008, p. 2.

24 *Sunday Tasmanian*, 11 May 2008, p. 2.

25 http://edition.cnn.com/2003/TACH/science/03/13/shark.study/, accessed 24 August 2007.

26 www.guardian.co.uk/world/2009/jan/13/australia-shark-attacks, accessed 24 August 2007.

27 Coppleson, V.M. (1958), *Shark Attack: A study of swimmers, surfers, skin divers, shipwreck survivors and sharks*, Sydney: Angus & Robertson, p. 68.

28 www.theage.com.au/articles/2004/07/11/1089484245921.html, accessed 24 August 2007.

29 www.news.com.au/story0, 10117, 17759065-1248,00.html, accessed 24 August 2007.

30 *The Australian*, 27 January 2008, accessed 28 January 2008.

31 *The Age*, 28 January 2008, p. 14. Letter written by Geoff Russell, St Morris, SA, accessed 29 January 2008.

32 www.ussindianapolis.org/woody.htm, accessed 25 August 2007. These are the words of Woody Eugene James.

33 *The Daily Telegraph*, Wednesday 9 August 2023, p. 25.

34 Ibid.

35 Ibid.

36 *The Daily Telegraph*, Saturday 5 August 2023, p. 8.

37 http://www.wipo.int/pctdb/en/wo.jsp?wo=2005085064, accessed 25 August 2007.

38 http://www.sharkshield.com/Content/Articles/View+Article/default. asp?id=1>, accessed 2 December 2008.

39 http://www.sharkcamo.com.au/, accessed 20 August 2007.

40 Coad, Brian W. and Papahn, Frough, op. cit. [see endnote 9 of this chapter].

41 https://ocean-guardian.com/, accessed 9 August 2023.

42 http://www.nswseakayaker.asn.au/magazine/50/bullshark.htm, accessed 25 August 2007.

2. Fathoming the Shark: *Evolution, Classification*

1 John Stainton's description of the footage of Steve Irwin's death as reported in *The Australian* newspaper, 5 September 2006, p. 1.

2 <www.theoi.com/Titan/Kirke.html>, accessed 10 September 2006. The epic poem that tells this story is known as the *Telegony* and was possibly composed in the sixth century BC. It followed the prediction of the blind seer Tieresias that the death of Odysseus would come 'out of the sea', which it did in the form of a stingray spine used as a spear tip, with which Telegonus accidentally killed Odysseus, his father.

3 Compagno, Leonardo J.V. (1987), 'Sharks and Their Relatives', in John D. Stevens (ed.), *Sharks,* Drummoyne: Golden Press, p. 62.

4 <www.elasmo-research.org/education/classification/what_makes.htm>, accessed 23 September 2006.

5 Hamlett, William C. (ed.) (1999), *Sharks, Skates and Rays: The biology of the elasmobranch fishes,* Baltimore MD: Johns Hopkins University Press, p. 2.

6 <www.shark.ch/Information/Evolution/index.html>, accessed 23 September 2006.

7 McCormick, Harold W. and Allen, Tom, with Young, Captain William E. (1964), *Shadows in the Sea: The sharks, skates and rays,* London: Sidgwick & Jackson, p. vii.

8 Helfman, Gene S., Collette, Bruce B. and Facey, Douglas E. (1999), *The Diversity of Fishes,* Malden, Mass.: Blackwell Science, p. 3.

9 Gross, M. Grant (1987), *Oceanography: A view of the earth,* 4th edn, Englewood Cliffs: Prentice-Hall, p. 33.

10 Harder, Ben (2002), 'Water for the rock: Did Earth's oceans come from the heavens?' in *Science News,* vol. 161, no. 12, p. 184.

11 <www.ga.gov.au/ausgeonews/ausgeonews200512/timescales.jsp>, accessed 25 May 2008.

12 Alexander, R. McN. (1974), *Functional Design in Fishes,* 3rd edn, London: Hutchinson University Library, pp. 19, 22.

13 <www.elasmo-research.org/education/evolution/golden_age.htm>, accessed 10 June 2007.

14 <www.bbc.co.uk/science/seamonsters/factfiles/stethacanthus.shtml>, accessed 10 June 2007.

15 <www.nhm.ac.uk/nature-online/life/reptiles-amphibians-fish/fathomseminar-jaws/session3/no-fish-fathomseminar-jaws-session3.html>, accessed 10 June 2007.

16 <www.geocities.com/ozraptor4/edestus.html>, accessed 10 June 2007.
17 <http://learn.amnh.org/courses/sharks_resource1.php>, accessed 10 June 2007.
18 <www.palaeos.com/Vertebrates/Units/070Chondrichthyes/070.100.html#Squatinactida>, accessed 7 December 2008.
19 <http://njfossils.net/hybodont.html>, accessed 15 July 2007.
20 Stead, David (1963), *Sharks and Rays of Australian Seas,* Sydney: Angus & Robertson, p. 46.
21 Compagno, Leonard J.V. (1999), 'Checklist of living elasmobranchs,' in Hamlett, William C. (ed.), *Sharks, Skates and Rays: The biology of the elasmobranch fishes,* Baltimore MD: Johns Hopkins University Press, pp. 471–98.
22 Souza, Shirley P. and Begossi, Alpina (2007), 'Whales, dolphins or fishes? The ethnotaxonomy of cetaceans in Sao Sebastiao, Brazil', *Journal of Ethnobiology and Ethnomedicine,* vol. 3, pp. 6, 8, <www.ethnobiomed.com/content/3/1/9>, accessed 8 April 2007.

3. Shark Biology: *Form and Function*

1 Shark researchers R. Aidan Martin and Anne Martin, quoted in <www.naturalhistorymag.com/1006/1006_feature.html>, accessed 10 January 2008.
2 Hamlett, William C. (ed.) (1999), *Sharks, Skates and Rays: The biology of the elasmobranch fishes,* Baltimore MD: Johns Hopkins University Press, p. 48.
3 ibid., p. 91.
4 Stevens, John D. (ed.) (1987), *Sharks,* Drummoyne: Golden Press, p. 51.
5 <www.elasmo-research.org/education/white_shark/muscles_jaw.htm>, accessed 13 June 2007.
6 <www.elasmo-research.org/education/white_shark/fins.htm>, accessed 13 June 2007.
7 Alexander, R. McN. (1974), *Functional Design in Fishes,* 3rd edn, London: Hutchinson University Library, p. 112.
8 Bright, Michael, 'Jaws: The natural history of sharks', Seminar Session 1, Natural History Museum, available online at <www.fathom.com/course/21701777/index.html>, accessed 26 December 2005.

9 Mojetta, Angelo (1997), *Sharks: History and biology of the lords of the sea,* Shrewsbury: Swan Hill Press, p. 54.
10 Dean, Mason N., Wilga, Cheryl D., Summers, Adam P. (2005), 'Eating without hands or tongue: Specialization, elaboration and the evolution of prey processing mechanisms in cartilaginous fishes', in *Biology Letters,* vol. 1, p. 357.
11 Pennisi, Elizabeth (2004), 'Shark flexes its teeth for tough meals', *Science,* vol. 303, no. 5660, p. 950.
12 <www.nature.com/nature/journal/v404/n6778/full/404566a0.html>, accessed 15 June 2008.

4. The Way of the Shark Roads: *Sharks and Indigenous Societies*

1 Köhnke, Glenys (1974), *The Shark Callers: An ancient fishing tradition of New Ireland, Papua New Guinea,* Boroko: Yumi Press, p. 16.
2 Torben, C. Rick, Erlandson, Jon M., Glassow, Michael A., Moss, Madonna L. (2002), 'Evaluating the economic significance of sharks, skates and rays (Elasmobranchs) in prehistoric economies', *Journal of Archaeological Science,* vol. 29, pp. 111–22 (quote taken from p. 112).
3 ibid., p. 114.
4 ibid., p. 116.
5 ibid., p. 117.
6 Gracey, Michael (2000), 'Historical, cultural, political, and social influences on dietary patterns and nutrition in Australian Aboriginal children', *The American Journal of Clinical Nutrition,* vol. 72, no. 5, 1361S–1367s, <www.ajcn.org/cgi/content/full/72/5/1361s>, accessed 16 January 2007.
7 McDavitt, Matthew T., (2005), 'The cultural significance of sharks and rays in Aboriginal societies across Australia's Top End', information sheet prepared for Shark Bay Seaweek 2005 [Natural Heritage Trust, Australian Government Department of Agriculture, Fisheries and Forestry, Marine Education Society of Australia], p. 3, <www.mesa.edu.au/seaweek2005/pdf_senior/is08.pdf>, accessed 1 September 2006.
8 Matthews, R.H. (1910), 'Reminiscences of Maori life fifty years ago', *Transactions and Proceedings of the Royal Society of New Zealand 1868–1961,* vol. 43, pp. 602–3.

9 <www.fao.org/DOCREP/005/X3690E/x3690e0q>, accessed 8 April 2007.

10 <icecook,blogspot.com/2006/01/how-to-prepare-hakarl-rotten-or-cured_17.html>, accessed 8 April 2007.

11 Gudger, E.W. (1927), 'Wooden hooks used for catching sharks and *Ruvettus* in the South Seas; a study of their variation and distribution', in *Anthropological Papers of The American Museum of Natural History,* New York City: The American Museum of Natural History, pp. 201–359.

12 ibid., p. 210.

13 ibid., p. 209.

14 Gardiner, J. Stanley (1898), 'The natives of Rotuma', in *Journal of the Royal Anthropological Institute,* vol. 27, p. 425, <www2.hawaii.edu/oceanic/rotuma/os/hanua.html>, accessed 4 December 2008.

15 Gudger, op. cit., p. 218.

16 ibid., p. 238.

17 ibid., p. 220, quoting Otto Finsch.

18 In some parts of Hawaii, ground-up plants (the '*auhuhu* and the *akia*) produced a substance that, when diluted in a tidal pool or underwater cave, stupefied and also killed fish. See <www.primitiveways.com/fish_toxins.html>, accessed 4 December 2008.

19 Köhnke, op. cit., pp. 9, 11, 13.

20 ibid., pp. 46–8.

21 Ellis, William (1833), *Polynesian Researches, During a Residence of Nearly Eight Years in the Society and Sandwich Islands,* New York: J. & J. Harper, quoted in Harold W. McCormick and Tom Allen with Captain William E. Young (1963), *Shadows in the Sea: The sharks, skates and rays,* London: Sidgwick & Jackson, pp. 192–3.

22 Gonzalez, Manoel M.B. (2005), 'Use of *Pristis* spp. (Elasmobranchii: Pristidae) by hunter–gatherers on the coast of Sao Paulo, Brazil', in *Neotropical Ichthyology* 3(3), pp. 421–6.

23 McDavitt, op. cit., p. 3.

24 ibid.

25 Khan, Kate (2003), *Catalogue of the Roth Collection of Aboriginal Artefacts from North Queensland,* vol. 3, Sydney: Rodin Print Pty Ltd, p. 82.

26 The *pahu* is made from a hollowed coconut trunk, with an attached rope of sennit, its use once restricted to religious ceremonies.

27 Kozuch, Laura (1993), *Sharks and Shark Products in Prehistoric South Florida,* Monograph Number 2, Institute of Archaeology and Paleoenvironmental Studies, Gainesville: University of Florida, p. 32.

28 <http://darwin-online.org.uk/content/search-results?freetext=snuff>, accessed 30 October 2006.

29 McDavitt, op. cit., p. 4.

30 <www.delange.org/TemMayor2/TemMayor2.htm>, accessed 30 October 2006.

31 McDavitt, Matthew T. (2002), 'Cipactli's sword, Tlaltecuhtli's teeth: deciphering the sawfish & shark offerings in the Aztec Great Temple', *Shark News 14,* <www.flmnh.ufl.edu/fish/organizations/ssg/sharknews/sn14/shark14news7.htm>, accessed 1 July 2007.

32 <www.tongatapu.net.to/lore/tonga/tonga104.htm>, accessed 1 November 2006.

33 Greene, Linda Wedel (1993), *A Cultural History of Three Traditional Hawaiian Sites on the West Coast of Hawai'i Island,* Denver, United States Department of the Interior National Park Service, p. 3, <www.cr.nps.gov/history/online_books/kona/history1e.htm>, accessed 19 May 2007.

34 Holt, John Dominis (1993), *Recollections,* Honolulu: Kupa'a, <http://apdl.kcc.hawaii.edu/~oahu/stories/koolauloa/sharks.htm>, accessed 22 May 2007.

35 Robinson, George Augustus (1829), *Friendly Mission: The Tasmanian journals and papers of George Augustus Robinson, 1829–1834,* ed. N.J.B. Plomley, Hobart: Quintus Publishing, 2008, p. 1042.

36 Montgomery, Charles (2004), *The Shark God: Encounters with myth and magic in the southern Pacific,* London: Fourth Estate, pp. 171, 279.

37 <www.projetmuse.net/index.php?id=1147&L=1.&S=0>, accessed 25 May 2007; <www.metmuseum.org/explore/oracle/figures49.html>, p. 2, accessed 25 May 2007.

38 Köhnke, op. cit., pp. 10, 12.

5. 'This Straunge and Merueylous Fyshe': *Sharks and Europeans*

1 William Shakespeare, *Macbeth* (4.1.5–37), c. 1605, <http://shakespeare.mit.edu/macbeth/macbeth.4.1.html>, accessed 4 December 2008.

2 Weis, H. Anne (2000), 'Odysseus at Sperlonga: Hellenistic hero or Roman foil?', in de Grummond, Nancy T. and Ridgway, Brunilde S., *From Pergamon to Sperlonga: Sculpture and context,* Berkeley: University of California Press, pp. 155–6.

3 <www.espr-archeologia.it/articles/103/The-Hellenistic-cooking>, accessed 27 May 2007.

4 Rapp, Albert (1955), 'The father of western gastronomy', in *The Classical Journal,* vol. 51, no. 1, p. 43.

5 Pliny the Elder (1940), *Natural History—With an English translation in ten volumes, Volume III,* trans. by H. Rackham, London: William Heinemann Ltd, repr. 1947, pp. 261, 269.

6 Andrewe, Lawrens, *The Noble Lyfe & Nature of Man, Of Bestes, Serpentys, Fowles & Fisshes y Be Moste Knowen,* translated by James L. Matterer and published online in 2000 as *Fantastic Fish of the Middle Ages.* See <www.godecookery.com/ffissh/ffissh.htm>, accessed 10 October 2006.

7 ibid.

8 Ashton, John (1890), *Curious Creatures in Zoology,* London: J.C. Nimmo, e-published (2000) by Armant Biological Press, p. 154.

9 de Casteau, Lancelot (1604), *L'Ouverture de Cuisine [Opening the Kitchen],* Liège: Leonard Street, trans. by Daniel Myers. See also <www.medievalcookery.com/notes/ouverture.shtm>, accessed 1 March 2007.

10 Roberts, J.M. (1980), *The Pelican History of the World,* Harmondsworth: Penguin, p. 643.

11 Boorstin, Daniel (1983), *The Discoverers,* Harmondsworth: Penguin, p. 386.

12 <www.etymonline.com/index.php?search=shark&searchmode=none>, accessed 4 June 2007. There are variations on the authorship and wording of this frequently quoted phrase.

13 *The Compact Edition of the Oxford English Dictionary, Volume II* (1979), London, Book Club Associates, p. 2771; <www.flmh.ufl.edu/fish/sharks/innews/selach2005.html>, accessed 1 December

2006; <www.etymonline.com/index.php?search=shark&searchmode=none>, accessed 4 June 2007.

14 *Compact OED,* ibid.

15 Castro, Jose I. (2002), 'On the origins of the Spanish word "tiburon", and the English word "shark"', *Environmental Biology of Fishes,* vol. 65, p. 250. See also Jones, Tom (1985), 'The *Xoc,* the *Sharke,* and the Sea Dogs: An historical encounter', in Robertson, M.G. and Fields, V.M., eds, Papers of the Fifth Palenque Round Table conference 1983, vol. vii, San Francisco: The Pre-Columbian Art Research Institute, 1983.

16 <http://homepage.mac.com/mollet/Av/Av1569.html>, accessed 4 June 2007. Image courtesy of Richard Lord, Sealord Photography <www.sealordphotography.net>.

17 From John Earle, *Microcosmography,* first published in 1628, quoted in Miehl, Dieter, Stock, Angela and Zwierlein, Anne-Julia (eds) (2004), *Plotting Early Modern London: New essays on Jacobean city comedy,* Aldershot: Ashgate, p. 90.

18 Roberts, op. cit., p. 643.

19 Acts 28:1–6, <www.iaudiobible.com/Online_Bible/NT/Acts/Chapter_28.htm>.

20 Cutler, Alan (2003), *The Seashell on the Mountaintop: A story of science, sainthood, and the humble genius who discovered a new history of the earth,* London: Heinemann, pp. 6–7.

21 *Elementary Mylogical Specimens* (1669) and *A Dissertation Concerning Solids Naturally Contained Within Solids* (1671).

22 Dampier, William (1939), *A Voyage to New Holland,* 1703 and 1709, Edited, with Introduction, Notes and Illustrative Documents, by James A. Williamson, London: The Argonaut Press, pp. 78, 107.

23 Willemsen, Mathieu (1997), 'Shagreen in western Europe: Its use and manufacture in the seventeenth and eighteenth centuries, *Apollo: The international magazine of arts,* vol. 419, p. 37.

24 Plomley, N.J.B. (ed.) (2008), *Friendly Mission: The Tasmanian journals and papers of George Augustus Robinson 1829–1831,* Hobart: Quintus Publishing, p. 111.

25 <www.darwin-literature.com/The_Voyage_Of_The_Beagle/2.html>, accessed 2 March 2007.

6. 'An Incredibly Bountiful Crop': *Shark Exploitation*

1 McCormick, Harold W. and Allen, Tom, with Young, Captain William E. (1963), *Shadows in the Sea: The sharks, skates and rays*, London: Sidgwick & Jackson, p. 169.

2 Lack, Mary and Sant, Glenn (2006), 'World shark catch, production & trade 1990–2003', TRAFFIC Oceania [supported by the] Australian Government Department of the Environment and Heritage, p. 4.

3 Kyne, Peter M. and Simpfendorfer, Colin A. (2007), 'A collation and summarization of available data on deepwater chondrichthyans: Biodiversity, life history and fisheries. A report prepared by the IUCN SSC Shark Specialist Group for the Marine Conservation Biology Institute', Bellevue: Washington, p. 88.

4 https://www.traffic.org/site/assets/files/12427/top-20-sharks-web-1.pdf, quoting FAO FishStat 2019, accessed 10 August 2023.

5 Ibid.

6 https://www.fao.org/3/cc0461en/online/sofia/2022/world-fisheries-aquaculture-production.html, accessed 10 August 2023.

7 Mitchell, Selina and Wilson, Lauren (2007), 'Fish bans raise food poison risk', Weekend Australian, 4–5 August, p. 1.

8 ibid. Mitchell and Wilson quoting Dr Peter Collignon, Infectious Diseases Unit, Canberra Hospital.

9 Gianluigi, Ferri, *et. al.*, 'Antiobiotic Resistance in the Finfish Aquaculture Industry: A Review', in *Antiobotics (Basel)*, November 2022, 11(11), p. 1574, published online by the National Library of Medicine, Bethesda, MD, https://www.ncbi.nlm.nih.gov/pmc/articles/PMC9686606/, accessed 10 August 2023.

10 <http://www.wildsingapore.com/news/20050910/051003-3.htm, accessed 14 September 2008.

11 http://www.fishonline.org/caught_at_sea/methods/, accessed 1 June 2007.

12 https://www.worldwildlife.org/species/shark, accessed 10 August 2023.

13 Murphy, Damien (2006), 'Flaky renaming fails to net shark diners', *Sydney Morning Herald*, 14 January, accessed 1 August 2008.

14 Vannuccini, Stefania (1999), 'Shark Utilization, Marketing and Trade', FAO Fisheries Technical Paper 389, Rome: Food and Agricultural Organization of the United Nations, see www.fao.org/docrep/005/x3690e/x3690e00.htm., accessed 15 April 2008.

15 Sample, Ian (2006), 'Sharks pay price for human tastes', *Guardian Weekly*, 8–14 September, p. 19.

16 Clarke, S. (2004), Shark Product Trade in Hong Kong and Mainland China and Implementation of the CITES Shark Listings, Hong Kong: TRAFFIC East Asia, p. 15.

17 https://www.worldwildlife.org/species/shark, accessed 10 August 2023.

18 https://www.theguardian.com/environment/2022/mar/02/european-countries-dominate-half-asian-shark-fin-trade-ifaw-report-reveals, accessed 10 August.

19 Taxin,Amy(2003),'"Finning"threatensGalapagossharks',CDNN Eco News, www.cdnn.info/eco/e031020a/e031020a.html, accessed 27 December 2005. See also www.flmnh.ufl.edu/fish/sharks/innews/galapagos2003.htm, accessed 10 December 2008.

20 Clarke, Shelley C., et al. (2006), 'Global estimates of shark catches using trade records from commercial markets', Ecology Letters 9, pp. 1120 and 1122.

21 Ibid., p. 1122.

22 www.lifetimehealth.com/squalene.asp, accessed 4 August 2007. Lifetime Health is an Australian health website. Its squalene products are marketed with the official Australian Made logo and it states that its oil is from bycatch, not targeted species.

23 Vannuccini, Stefania, op. cit., Section 6.4.2.

24 Gilman, Eric, et al. (2007), Shark depredation and unwanted bycatch in pelagic longline fisheries: Industry practices and attitudes, and shark avoidance strategies 2007, Honolulu: Western Pacific Regional Fishery Management Council, p. viii.

25 Bevilaqua, Simon (2006), 'Submerged treasures', *Sunday Tasmanian,* 19 February, p. 11. Bevilaqua quoted Humane Society International spokesman Michael Kennedy.

26 <www.fish.wa.gov.au/sec/rec.index>, accessed 1 August 2007.

27 White, T.W., et al. (2006), Economically Important Sharks and Rays of Indonesia, Canberra: Australian Centre for International Agricultural Research, p. 1.

7. Shark Conservation: *Problems, Solutions*

1 Casey, Susan (2005), The Devil's Teeth: A true story of obsession and survival among America's great white sharks, New York: Henry Holt & Co., p. 3.

2 Richardson, Michael (2005), 'The fished-out planet', The Age, 27 December, p. 11.

3 Ibid.

4 www.bite-back.com/sharks, accessed 30 December 2005. Bite-Back is a shark and marine conservation organisation.

5 Compagno, Leonardo J.V. (2000), 'Sharks, fisheries and biodiversity', paper presented at the Pacific Fisheries Coalition Shark Conference 2000, Honolulu, Hawaii, 21–24 February, available online at www.pacfish.org/sharkcon/documents/compagno.html, accessed 17 June 2007.

6 www.sharkinfo.ch/SI2_99e/requiem.html, accessed 6 August 2007. The Shark Foundation is a Zurich-based shark conservation organisation.

7 https://sharkangels.org/history/, accessed 12 August 2023.

8 'Shark exploitation in Ghana hastens global collapse', ENS Release 27 August 2001, available online at www.flmnh.ufl.edu/fish/sharks/innews/sharkexp2001.html, accessed 6 August 2007.

9 Baum, Julia K., et al. (2003), 'Collapse and conservation of shark populations in the Northwest Atlantic', Science, vol. 299, no. 5605, p. 390.

10 www.diveoz.com.au/gns/gns-threats.asp, accessed 6 August 2007.

11 https://www.iucnssg.org/iucnredlist.html, accessed 10 August 2023.

12 https://www.dcceew.gov.au/environment/biodiversity/threatened/action-plan/priority-fish/grey-nurse-shark, accessed 10 August 2023.

13 IUCN News Release, 22 February 2007, available online at www.iucn.org/en/news/archive/2007/02/22_pr_sharks.htm, accessed 1 August 2007.

14 https://www.iucnssg.org/news/new-global-study-finds-unprecedented-shark-and-ray-extinction-risk, accessed 11 August 2023.

15 The Shark Foundation, www.sharkinfo.ch/S14_))e/list.html, accessed 6 August 2007, accessed 6 August 2007.

16 https://www.iucn.org/story/202209/new-hope-conservation-sharks-rays-and-chimaeras-important-shark-and-ray-areas-isras#:~:text=The%20most%20recent%20global%20IUCN, et%20al.%2C%202021), accessed 10 August 2023.

17 Pepperell, Julian (2005), 'Release of sharks in recreational fisheries', Shark Bay Seaweek 2005–Save Our Sharks, <www.mesa.edu.au/seaweek2005>, accessed 5 August 2007.

18 www.asoc.org/what_seamounts.htm, accessed 10 August 2007.

19 Ibid.

20 Kyne, Peter M. and Simpfendorfer, Colin A. (2007), 'A Collation and Summarization of Available Data on Deepwater Chondrichthyans: Biodiversity, life history and fisheries', Report prepared by the IUCN SSC Shark Specialist Group for the Marine Conservation Biology Institute, p. 4, available at <http://flmnh.ufl.edu/fish/organizations/ssg/deepchondreport.pdf>, accessed 11 August 2007.

21 Stevens, John D., personal communication with the author, 3 September 2008.

22 Bonfil, Ramon, et al. (2005), 'Transoceanic migration, spatial dynamics, and population linkages of white sharks', Science, vol. 310, p. 100.

23 The areas concerned are: Northeast Atlantic; Mediterranean; West Africa; Sub-equatorial Africa; Northwest Atlantic; Central America and Caribbean; Indian Ocean; Australasia and Oceania; Southeast Asia; Northwest Pacific; and Northeast Pacific. See also www.flmnh.ufl.edu/fish/organizations/ssg/ssg/htm.

24 http://www.fao.org/fi/website/FIRetrieveAction.do?dom=org&xml=ipoa_sharks.xml&xp_nav=2>, accessed 11 August 2007.

25 www.flmnh.ufl.edu/fish/organizations/ssg/cites.htm, accessed 11 August 2007.

26 Rose, Cassandra (2002), 'Recreational shark catch in Australia', in The Australian Shark Assessment Report for the Australian National Plan of Action for the Conservation and Management of Sharks, Canberra: Commonwealth Department of Agriculture, Fisheries and Forestry, p. 1.

27 Shark Advisory Group and Lack, Mary (2004), *National Plan of Action for the Conservation and Management of Sharks (Shark-plan)*, Canberra: Department of Agriculture, Fisheries and Forestry, p. 3.

28 https://www.agriculture.gov.au/agriculture-land/fisheries/environment/sharks, accessed 12 August 2023.

29 reide, N.R.J., et al. (2007), European Shark Fisheries: A preliminary investigation into fisheries, conversion factors, trade products, markets and management measures, Newbury: European Elasmobranch Association, p. 53.

30 https://elasmo.org/about/, accessed 12 August 2023.

31 Ibid, accessed 14 August 2007.

32 https://elasmo.org/wp-content/uploads/2022/04/Report-on-the-AES-Global-Wedgefish-and-Guitarfish-Symposium.pdf, accessed 12 August 2023.

33 http://ftp.marine.csiro.au/tagging/whitesharks/taggedsharks.html>, accessed 15 July 2007.

34 https://www.sharksmart.com.au/research/shark-tagging/, accessed 12 August 2023.

35 www.oceaniasharks.org.au/About-Us.aspx, accessed 15 August 2007.

36 http://www.oceaniasharks.org.au/Media,/news/and/events/News/stories/26/9/06/News/May/25/SP/2007.html, accessed 15 August 2007.

37 http://www.flmnh.ufl.edu/fish/sharks/nsrc/NSRC.htm, accessed 14 August 2007.

38 http://www.flmnh.ufl.edu/fish/sharks/nsrc/featproj05.htm, accessed 14 August 2007.

39 https://www.floridamuseum.ufl.edu/sharks/about/, accessed 12 August 2023.

40 https://www.shark.co.za/Pages/AboutUs-Overview accessed 12 August 2023.

41 https://www.shark.co.za/Pages/ProtectionSharks-NetsDrumlines, accessed 12 August 2023.

42 https://www.sharkconservationfund.org/the-crisis/, accessed 12 August 2023.

43 https://www.biminisharklab.com/about, accessed 12 August 2023.

44 https://www.sharks.org/mission-work, accessed 12 August 2023.

45 Ibid.

46 https://wildoceans.org/our-mission/, accessed 13 August 2023.

47 https://wildoceans.org/respite-for-mako/, accessed 13 August 2023.

48 https://www.padi.com/aware, accessed 13 August 2023.
49 https://www.marineconservation.org.au/about/, accessed 13 August 2023.
50 Ibid.
51 Ibid.
52 https://www.sharktrust.org/Handlers/Download.ashx?IDMF=333a9a7c-7512-46e5-9dbb-6a8b6f7882a0, accessed 13 August 2023.
53 https://oceanconservancy.org/about/history/, accessed 13 August 2023.
54 https://oceanconservancy.org/programs/, accessed 13 August 2023.
55 https://shark.swiss/foundation/, accessed 13 August 2023.
56 https://shark.swiss/projects, accessed 13 August 2023.
57 https://www.bite-back.com/what-we-do/?v=6cc98ba2045f, accessed 13 August 2023.
58 https://www.bite-back.com/what-we-do/?v=6cc98ba2045f, accessed 13 August 2023.
59 Ibid.
60 Shark Angels - Turning Fear Into Fascination, accessed 14 August 2023.
61 https://www.mcsuk.org/about-us/, accessed 13 August 2023.
62 http://www.mcsuk.org/, accessed 13 September 2008.
63 https://oceana.org/about/, accessed 13 August 2023.
64 https://www.imas.utas.edu.au/research, accessed 13 August 2023.
65 https://www.theguardian.com/environment/2023/may/16/scientists-warn-maugean-skate-tasmanias-thylacine-of-the-sea-one-extreme-weather-event-from-extinction, accessed 13 August 2023.
66 https://tasmanianinquirer.com.au/news/this-is-a-hard-conversation-endangered-maugean-skate-headed-for-extinction-in-a-decade-unless-macquarie-harbour-fish-farms-rested/, accessed 13 August 2023.
67 Personal email communication with the author, 24 August 2023.

8. Sharks and Creativity: *Visions of Hunter and Hunted*

1 Steven Spielberg, quoted in *Weekend Australian Magazine*, 26–27 April 2008, p. 19.

2 <www.metmuseum.org/explore/oracle/figures49.html>, accessed 26 May 2007.

3 <www.strangescience.net/stsea2.htm>, accessed 5 December 2008.

4 Rishel, Joseph J. (1982), 'A Lyonnais flower piece by Antoine Berjon (1754–1843)', *Philadelphia Museum of Art Bulletin,* vol. 78, no. 336, pp. 20, 23.

5 ibid.

6 Masur, Louis P. (1994), 'Reading *Watson and the Shark', The New England Quarterly,* vol. 67, no. 3, p. 427.

7 Boime, Albert (1989), 'Blacks in shark-infested waters: Visual encodings of racism in Copley and Homer', *Smithsonian Studies in American Art,* vol. 3, no. 1, p. 21.

8 Staiti, Paul (2001), 'Winslow Homer and the drama of thermodynamics', *American Art,* vol. 15, no. 1, pp. 24, 30–1.

9 <http://arts.guardian.co.uk/news/story/0,,1392473,00.html#article_continue>, accessed 30 August 2007.

10 Kent, Sarah (1994), *Shark Infested Waters: The Saatchi Collection of British art in the 90s,* London: Zwemmer, p. 35.

11 Kent, op. cit., p. 37.

12 ibid.

13 ibid.

14 Baker, Steve, *The Postmodern Animal,* London: Reaktion Books, 2000, p. 86.

15 <www.radio.cz/en/article/75583>, accessed 1 August 2008.

16 Baker, op. cit., p. 12.

17 <www.headington.org.uk/history/misc/shark>, accessed 21 July 2007.

18 ibid.

19 <www.interiordesign.net/id_article/CA306192/id?stt=001>, accessed 13 September 2008.

20 Kingston, W.H.G. (William Henry Giles) (1876), *The Three Lieutenants,* London: Griffith & Farran, pp. 150–1, 152.

21 Melville, Herman (1851), *Moby-Dick,* New York: Harper & Brothers, ch. 66.

22 Poe, Edgar Allan (1838), *The Narrative of A. Gordon Pym,* New York: Harper & Brothers, ch. xiii.

23 Hemingway, Ernest (1952), *The Old Man and the Sea,* New York: Scribner, various pages.

24 <www.peterbenchley.com>, accessed 25 August 2007.
25 ibid.
26 ibid.
27 Quirke, Antonia (2002), *Jaws* [BFI Modern Classics series], London: British Film Institute, p. 67.
28 Shudder's 'Sharksploitation' Gives the Subgenre the Respect It Deserves (collider.com), accessed 14 August 2023, accessed 14 August 2023.
29 Ibid.
30 Ibid.
31 Ibid.

9. 'Creatures of Extremes': *Descriptions of Sharks, Skates, Rays and Chimaeras*

1 <www.elasmo-research.org/education/shark_profiles/squaliformes.htm>, accessed 1 November 2006; <http://zipcodezoo.com/Animals/S/Squaliolus_laticaudus/>, accessed 7 December 2008; <www.flmnh.ufl.edu/fish/Gallery/Descript/PortugueseShark/PortugueseShark.html>, accessed 8 December 2008.
2 <http://sharks-med.netfirms.com/med/anglrough.htm>, accessed 1 June 2007.
3 Last, P.R. and Stevens, J.D. (1994), *Sharks and Rays of Australia,* Hobart: CSIRO Australia, p. 98.
4 Stead, David G., *Sharks and Rays of Australian Seas,* Sydney: Angus & Robertson, 1963, p. 121.
5 *UBC Reports,* vol. 52, no. 12, 7 December 2006.
6 ibid.
7 Harvey-Clark, Chris J. and Gallant, Jeffrey J. (2005), 'Vision and its relationship to novel behaviour in St. Lawrence River Greenland Sharks, *Somniosus microcephalus', The Canadian Field-Naturalist,* vol. 119, no. 3, pp. 355–9.
8 <www.iucnredlist.org/search/details.php/41801/all>, accessed 21 June 2007.
9 Stevens, John D., consulting editor, *Sharks,* Drummoyne: Golden Press, 1987, p. 21.
10 Last, P.R. and Stevens, J.D., op. cit., p. 106.
11 <www.flmnh.ufl.edu/fish/organizations/ssg/redlist.htm>, accessed 17 June 2007.

12 Heemstra, Phillip C., *et al.* (2006), 'Interactions of fishes with particular reference to coelacanths in the canyons at Sodwana Bay and the St Lucia Marine Protected Area of South Africa', *South African Journal of Science,* September, p. 461.

13 Summers, Adam P. and Koob, Thomas J. (1997), 'A biographical sketch of Samuel Walton Garman', pp. vii–viii, <http://biomechanics.bio.uci.edu/_media/pdf_papers/plagbio.pdf>, accessed 17 June 2007.

14 <www.elasmo-esearch.org/education/topics/d_jurassic_shark.htm>, accessed 31 August 2008.

15 Larson, Shawn (2005), 'Sixgill shark (*Hexanchus griseus*) conservation ecology project update', in *Proceedings of the 2005 Puget Sound Georgia Basin Research Conference,* p. 1.

16 <www.elasmo-research.org/education/shark_profiles/carcharhiniformes.htm>, accessed 4 December 2008.

17 Last and Stevens, op. cit., p. 259.

18 <www.marine.csiro.au/media/03releases/17jan03.htm>, accessed 1 December 2008.

19 Whitney, Nicholas M., Pratt, Harold L. Jr and Carrier, Jeffrey C. (2004), 'Group courtship, mating behaviour and siphon sac function in the whitetip reef shark, *Triaenodon obesus*', *Animal Behaviour,* vol. 68, p. 1439.

20 <www.wetwebmedia.com/blacktipshark.htm>, accessed 10 June 2007.

21 Compagno, Leonard J.V. (1984), *Sharks of the World: An annotated and illustrated catalogue of shark species known to date, Part 2–Carcharhiniformes,* FAO Species Catalogue, vol. 4, part 2 (FAO Fisheries Synopsis No. 125), Rome: United Nations Development Programme, Food and Agricultural Organization of the United Nations, p. 461.

22 ibid.

23 <www.pewoceanscience.org/projects/Glovers_Reef_/intro.php?ID=54>, accessed 16 June 2007.

24 <www.qm.qld.gov.au/organisation/e_prints/mqm_51_2/51_2_Kyne-et-al.pdf>, accessed 28 June 2008.

25 <www.elasmo-research.org/education/white_shark/vision.htm>, accessed 5 December 2008.

26 <www.flmnh.ufl.edu/fish/Gallery/Descript/OceanicWT/Oceanic

WT.html>, accessed 30 June 2007.

27 <www.literature.org/authors/darwin-charles/the-voyage-of-the-beagle/chapter-01.html>, accessed 1 July 2007.

28 <http://gdl.cdlr.strath.ac.uk/scotia/gooant/gooant/0502.htm>, accessed 30 June 2007.

29 Lubbock, H.R. and Edwards, A.J. (1982), 'The shark population of Saint Paul's Rocks', *Copeia,* vol. 1982, no. 1, p. 223.

30 Lubbock, H.R. and Edwards, A.J. (1981), 'The fishes of Saint Paul's Rocks', *Journal of Fish Biology,* vol. 18, issue 2, pp. 135–57.

31 Feitoza, M., *et al.* (2003), 'Reef fishes of St Paul's Rocks: New records and notes on biology and zoogeography', *Aqua, Journal of Icthyology and Aquatic Biology,* vol. 7, no. 2, p. 67.

32 ibid.

33 <www.flmnh.ufl.edu/fish/Gallery/Descript/BlueShark/BlueShark.html>, accessed 30 June 2007.

34 <www.elasmo-research.org/education/shark_profiles/carcharhinidae.htm>, accessed 30 June 2007.

35 Carey, F.G. and Scharold, J.V. (1990), 'Movements of blue sharks (*Prionace glauca*) in depth and course', *Marine Biology,* vol. 106, p. 330.

36 Stevens, John D., personal communication to the author, 3 September 2008.

37 <www.abc.net.au/catalyst/stories/s903678.htm#>, accessed 5 December 2008.

38 <www.underwater.com.au/article.php/id/2103/>, accessed 2 July 2007. See also <www.projectaware.org>.

39 <www.adventuredivingsafaris.co.za/speciesinformationbullsharks.html>, accessed 5 May 2007.

40 <http://marinebio.org/species.asp?id=87>, accessed 4 July 2007.

41 <www.flmnh.ufl.edu/fish/Gallery/descript/Schammer/Scalloped-Hammerhead.html>, accessed 4 July 2007.

42 Last and Stevens, op. cit., pp. 270, 274.

43 Kajiura, Stephen M. and Holland, Kim N. (2002), 'Electroreception in juvenile scalloped hammerhead and sandbar sharks', *Journal of Experimental Biology,* vol. 205, p. 3609.

44 ibid.

45 Phillips, Kathryn (2002), 'How the shark got its hammer head', *Journal of Experimental Biology,* vol. 205, p. 2304.

46 Mojetta, Angelo (1997), *Sharks: History and biology of the lords of the sea,* Shrewsbury: Swan Hill Press, p. 42.
47 <http://news.nationalgeographic.com/news/2002/06/0606_020606_shark4.html>, accessed 6 July 2007.
48 Simpfendorfer, Colin (1992), 'Biology of tiger sharks (*Galeocerdo cuvier*) caught by the Queensland shark meshing program off Townsville, Australia', *The Australian Journal of Marine and Freshwater Research,* vol. 43, pp. 39–40.
49 <www.elasmo.com/frameMe.html?file=genera/cenozoic/sharks/galeocerdo.html&menu=bin/menu_genera-alt.html>, accessed 10 July 2007.
50 <www.fiu.edu/~heithaus/Reseach.htm>, accessed 1 June 2008.
51 <http://cat.inist.fr/?aModele=afficheN&cpsidt=13480355>, accessed 1 July 2007; Abstract of article by M.R. Heithaus, *et al.* (2002), 'Habitat use and foraging behavior of tiger sharks (*Galeocerdo cuvier*) in a seagrass ecosystem', *Marine Biology,* vol. 140, no. 2, pp. 237–48.
52 <www.practicalfishkeeping.co.uk/pfk/pages/item.php?news=1080>, accessed 16 June 2007.
53 <www.fishbase.org/summary/SpeciesSummary.php?id=825>, accessed 5 December 2008.
54 Last and Stevens, op. cit., p. 131.
55 <www.dpi.nsw.gov.au/research/areas/systems-research/aquatic-ecosystems/outputs/2007/883>, accessed 1 December 2008.
56 Huveneers, Charlie (2006), 'Redescription of two species of wobbegongs (Chondrichthyes: Orectolobidae) with elevation of *Orectolobus halei* Whitley 1940 to species level', *Zootaxa,* vol. 1284, p. 29.
57 <www.julianrocks.net/sharks/Orectolobus.htm>, accessed 10 June 2007.
58 <www.nigelmarshphotography.com/articles/wobbies.htm>, accessed 1 December 2008.
59 <www.elasmo-research.org/education/ecology/intertidal-nurse.htm>, accessed 20 June 2007.
60 Daley, R.K., *et al.* (2002), *Field Guide to Australian Sharks and Rays,* Collingwood: CSIRO Publishing, p. 32.
61 Dudgeon, Christine L. (2005), 'The ecology of the leopard shark *Stegostoma fasciatum*', in *Proceedings of the Inaugural Southern*

Queensland Elasmobranch Research Forum, Brisbane: University of Queensland, Faculty of Biological and Chemical Sciences, p. 3.

62 Stead, David G., *Sharks and Rays of Australian Seas,* Sydney: Angus & Robertson, 1963, p. 51.

63 <www.elasmo-research.org/education/topics/d_filter_feeding.htm>, accessed 22 July 2007.

64 *The Voyage of Governor Phillip to Botany Bay With an Account of the Establishment of the Colonies of Port Jackson and Norfolk Island,* 1789, Chapter XV.

65 <www.postmodern.com/~fi/sharkpics/ellis/goblin.htm>, accessed 2 May 2007.

66 <www.flmnh.ufl.edu/FISH/Gallery/Descript/GoblinShark/GoblinShark.html>, accessed 5 May 2007.

67 <www.sportsfish.com.au/pages/fishing/fish-saltwater/makoshark.html>, accessed 5 May 2007.

68 <http://ichthy.mlml.calstate.edu/ardizzone.htm>, accessed 6 May 2007.

69 <www.flmnh.ufl.edu/fish/Gallery/Descript/Porbeagle/Porbeagle.html>, accessed 10 December 2008.

70 <www.elasmo-research.org/education/shark_profiles/l_nasus.htm>, accessed 6 May 2007.

71 <http://marinebio.org/species.asp?id=92>, accessed 6 May 2007.

72 Last and Stevens, op. cit., p. 153.

73 <www.cheetah.org>, accessed 2 July 2007.

74 <www.geocities.com/yosemite/1133/thresher.html>, accessed 2 July 2007.

75 *The Mercury,* 10 July 2006, p. 18.

76 Steel, Rodney (1992), *Sharks of the World,* London: Blandford Press, p. 131.

77 Sims, David W., *et al.* (2000), 'Annual social behaviour of basking sharks associated with coastal front areas', in *Proceedings of the Royal Society B: Biological Sciences,* London: The Royal Society, vol. 267, pp. 1897–1904.

78 Sims, David W., *et al.* (2005), 'Habitat-specific normal and reverse diel vertical migration in the plankton-feeding basking shark', *Journal of Animal Ecology,* vol. 74, p. 760.

79 Shuker, Karl P.N., 'Bring me the head of the sea serpent', quoted in <www.baskingshark.org>, accessed 15 July 2007.

80 Douady, Christophe J., *et al.* (2003), 'Molecular phylogenetic evidence refuting the hypothesis of Batoidea (rays and skates) as derived sharks', *Molecular Phylogenetics and Evolution,* vol. 26, p. 215.

81 In September 2004 the Shark Specialist Workshop of the International Union for the Conservation of Nature (IUCN) conducted a batoid workshop in Cape Town, South Africa, attended by 29 chondrichthyan fish experts from fifteen countries.

82 Dean, Mason M., Wilga, Cheryl D. and Summers, Adam P. (2005), 'Eating without hands or tongue: specialization, elaboration and the evolution of prey processing mechanisms in cartilaginous fishes', in *Biology Letters,* vol. 1, no. 3, London: The Royal Society, p. 357.

83 Hamlett, William C. (1999), *Sharks, Skates, and Rays,* Baltimore MD: Johns Hopkins University Press, p. 29.

84 Daley, R.K., *et al.* (2002), *Field Guide to Australian Sharks and Rays,* Hobart: CSIRO Australia, p. 39.

85 Hamlett, op. cit., p. 485.

86 Last, Peter R. (2004), '*Rhinobatos sainsburyi* n. sp. and *Aptychotrema timorensis* n. sp.—Two new shovelnose rays (Batoidea: Rhinobatidae) from the Eastern Indian Ocean', *Records of the Australian Museum,* vol. 56, pp. 201–8.

87 <www.mbayaq.org/efc/living_species/default.asp?inhab=134>, accessed 20 July 2007.

88 <http://fishbase.org/Summary/SpeciesSummary.php?id=8729>, accessed 14 December 2008.

89 <www.flmnh.ufl.edu/fish/Meetings/abst2005a.html>, accessed 25 July 2007. The quotation is taken from the abstract of a paper delivered by Vicente Faria and Matthew McDavitt to the American Elasmobranch Society's 2005 Annual Meeting in Tampa, Florida. The paper is entitled 'Trying again two centuries later: An essay on the various species of sawfish'.

90 <www.floridasawfish.com>, accessed 20 June 2007.

91 Le Muséum national d'Histoire naturelle in Paris has inaugurated an Odontobase Project, run by Pascal P. Deynat.

92 <www.flmnh.ufl.edu/fish/Gallery/Descript/ThornySkate/ThornySkate.html>, accessed 26 July 2007.

93 Last and Stevens, op. cit., p. xx.

94 Skjaeraasen, J.E. and Bergstad, O.A. (2001), 'Notes on the distribution and length composition of *Raja lintea, R. fyllae, R. hyperborean and Bathyraja spinicauda* (Pisces: Rajidae) in the deep northeastern North Sea and on the slope of the eastern Norwegian Sea', *ICES Journal of Marine Science,* vol. 58, pp. 21–8.

95 <www.flmnh.ufl.edu/fish/Gallery/Descript/BarndoorSkate/BarndoorSkate.html>, accessed 1 August 2007.

96 <www.austmus.gov.au/fishes/fishfacts/fish/hendo.htm>, accessed 24 July 2007.

97 Last and Stevens, op. cit., p. 370.

98 Kyne, Peter M. and Simpfendorfer, Colin A. (2007), 'A collation and summarization of available data on deepwater chondrichthyans: Biodiversity, life history and fisheries. A report prepared by the IUCN SSC Shark Specialist Group for the Marine Conservation Biology Institute', Bellevue: Washington, p. 88.

99 Last, Peter R. and Compagno, Leonard J.V. (1999), 'Urolophidae: Stingarees', in Kent E. Carpenter and Volker H. Niem, *FAO Species Identification Guide for Fishery Purposes: The Living Marine Resources of the Western Central Pacific, Volume 3: Batoid Fishes, Chimaeras and Bony Fishes Part 1* (*Elopidae to Linophrynidae*), Rome: Food and Agricultural Organization of the United Nations, p. 1469.

100 Schwartz, Frank J. (2007), 'Tail spine characteristics of stingrays (Order Myliobatiformes) frequenting the FAO Fishing area (20°N 120°E—50°N 150°E) of the Northwest Pacific Ocean', *The Raffles Bulletin of Zoology,* supplement no. 14, pp. 121–30.

101 <http://animaldiversity.ummz.umich.edu/site/accounts/information/Hexatrygonidae.html>, accessed 31 August 2007.

102 <www.flmnh.ufl.edu/FISH/Gallery/Descript/CownoseRay/CownoseRay.html>, accessed 31 August 2007.

103 <www.flmnh.ufl.edu/fish/Gallery/Descript/MantaRay/MantaRay.html>, accessed 31 August 2007.

104 <www.flmnh.ufl.edu/fish/Gallery/Descript/spottedratfish/spottedratfish.html>, accessed 10 December 2008.

105 <www.fishnet.com.au/default.aspx?id=225&fishid=35>, accessed 10 June 2007.

INDEX

Milton Keynes UK
Ingram Content Group UK Ltd.
UKHW020611101123
432316UK00009B/46